GUIDELINES FOR

Safe Storage and Handling of High Toxic Hazard Materials

GUIDELINES FOR
Safe Storage and Handling of High Toxic Hazard Materials

Prepared by
Arthur D. Little, Inc.
Cambridge, Massachusetts

and

Richard LeVine, Ph.D.
Center for Chemical Process Safety

for

CENTER FOR CHEMICAL PROCESS SAFETY
of the
American Institute of Chemical Engineers
345 East 47th Street, New York, NY 10017

Library of Congress Cataloging-in-Publication Data

LeVine, Richard, (Date)
Guidelines for safe storage and handling of high toxic hazard materials.

Bibliography: p.
Includes index.
1. Hazardous substances — Safety measures. I. Arthur D. Little, Inc. II. American Institute of Chemical Engineers. Center for Chemical Process Safety. III. Title.
T55.3.H3L48 1987 604.7 87-26918
ISBN 0-8169-0400-6

CONTENTS

LIST OF TABLES AND FIGURES

TABLES

FIGURES

PREFACE

The American Institute of Chemical Engineers (AIChE) has a 30-year history of involvement with process safety and loss control for chemical and petrochemical plants. Through its strong ties with process designers, builders, and operators, safety professionals, and academia, the AIChE has enhanced communication and fostered improvement in the high safety standards of the industry. Their publications and symposia have become an information resource for the engineering profession on the causes of accidents and means of prevention.

Early in 1985, the AIChE established the Center for Chemical Process Safety (CCPS) to serve as a focus for a continuing program on process safety. The first AIChE-CCPS project was the preparation of "Guidelines for Hazard Evaluation Procedures." Since then the Center has initiated several other projects. One of the 1986-1987 projects was for work that led to the publication of this document, "Guidelines for Safe Storage and Handling of High Toxic Hazard Materials."

The safe storage and handling of high toxic hazard materials (HTHMs) require a sound and responsible management philosophy together with a combination of superior siting, design, fabrication, erection, inspection, monitoring, maintenance, operation, and management of such facilities. These elements are necessary parts of a reliable system to prevent equipment or human failures that might lead to the release of -- and possible acute exposure of the public or on-site personnel to -- HTHMs. These Guidelines deal with all these elements, including references for mitigating the impacts of such releases should they occur.

These Guidelines are a technical document. They are intended for use by engineers and other persons familiar with the manufacture and use of chemicals. They include discussion of some of the current industry practices for controlling toxic hazards both for existing facilities and for plants presently being designed. They are not a standard and make no attempt to cover all the legal requirements that may relate to the construction and operation of facilities for the storage and handling of HTHMs.

Meeting such legal requirements is a minimum basis for design and operation of all facilities. These Guidelines highlight and supplement those basic requirements that are particularly important to the safe storage and handling of HTHMs. Thus, they should be applied with engineering judgment as well as a knowledge of the hazards and properties of each particular toxic chemical.

The risk of a release of a highly toxic, unreactive material caused by the catastrophic structural failure of a well-designed and maintained bulk storage vessel can be reduced to very low levels. There may be a much greater likelihood and, thus, a greater risk caused by the failure of the attached equipment and piping, or by human operating error. The risk may be even higher if the material being stored is also reactive and/or flammable.

Existing facilities may have been designed and constructed to earlier versions of codes and standards and thus may not fully reflect current practices. When major modifications or additions are made to older facilities, the new portions should meet current design practices for new facilities. However, it is the responsibility of management to decide whether additional safety-related design changes in older facilities are necessary and warranted. Nevertheless, the management of existing facilities for the storage and handling of HTHMs should apply current standards and safety practices to their operating, maintenance, management, and emergency procedures and should also reassess safety monitoring and control systems to see whether enhancement of such systems is needed to meet current levels of good practice.

These Guidelines focus primarily on good design and operating practices for the storage and handling of HTHMs. AIChE-CCPS is preparing a family of guideline documents that will address different aspects of general process safety. In addition to "Guidelines for Hazard Evaluation Procedures," "Guidelines for Use of Vapor Cloud Dispersion Models" have been published. These give more information on quantifying dispersion of toxic vapor clouds than is included herein. Another publication that is scheduled for 1988 is "Guidelines for Vapor Release Mitigation." That volume covers mitigation after a release occurs, while this volume focuses on prevention of releases.

"Guidelines for Chemical Process Quantitative Risk Assessment Procedures" are scheduled for publication in 1988. That volume will give guidance on determining the level of risk from a facility, which may be used in determining the extent of protection that is justified.

The authors appreciate the opportunity to prepare these Guidelines for the AIChE-CCPS. In particular, we are grateful to the members of the AIChE-CCPS Task Force, who served on the subcommittee that interacted with Arthur D. Little, Inc. in the preparation of these Guidelines. They included

Mr. Robert A. Smith, Subcommittee Chairman
The Dow Chemical Company

Mr. Richard E. Munson
E.I. DuPont de Nemours & Company, Inc.

Dr. Dennis E. Wade
Monsanto Company

"Guidelines for Safe Storage and Handling of High Toxic Hazard Materials" were prepared by an Arthur D. Little, Inc. (ADL) team led by Dr. Paul A. Croce. An original draft of this report was prepared by Dr. Richard LeVine, of AIChE-CCPS staff, who also participated in the final review of the document. Russell G. Hill provided AIChE-CCPS staff support. Major contributors to the document from ADL included David L. Farrell, John H. Hagopian, Henry Ozog, and R. Peter Stickles. Members of the Arthur D. Little review panel included Stanley E. Dale, Dr. Elisabeth M. Drake, and R. Scott Stricoff. Editing was by Dana Pierce, and typing and revisions were by our dedicated secretaries, Rose Scali and Josephine Kiguli.

AIChE acknowledges the generosity of the Dow Chemical Company for permission to include The Chemical Exposure Index (Appendix B).

GLOSSARY

Accident, accident sequence	A specific unplanned event or sequence of events with a specific undesirable consequence.
Acute exposure	A short-term or rare exposure to a toxic agent in a single episode that is unlikely to recur.
Acute hazard	The potential for injury or damage to occur as a result of an instantaneous or short duration exposure to the effects of an undesirable event (e.g., an explosion with the potential for causing damage and injury).
Category M	A fluid service in which the potential for personnel exposure is judged to be significant, and in which a single exposure to a very small quantity of a toxic fluid, caused by leakage, can produce serious irreversible harm to persons upon breathing or bodily contact, even when prompt restorative measures are taken (ASME B31.3).
Chronic exposure	A frequent or continuous exposure to a toxic agent over an unspecified, but generally lengthy, period of time.
Chronic hazard	The potential for injury or damage to occur as a result of prolonged exposure to an undesirable condition (e.g., smoking, with the potential for causing lung cancer).

Consequence	The direct result of a hazardous event, usually expressed in safety assessments in terms of the magnitude of the area over which these impacts occur.
Design basis incident (DBI) (or maximum credible incident)	A serious incident that has some small likelihood of occurring during the lifetime of a facility but consequences of which (resulting hazard zones) are used in siting, plant layout, and/or emergency planning decisions. (See discussions in Chapter 2.)
Emergency exposure guidance limits (EEGLs)	Maximum concentration levels that provide guidance in advance planning for management of emergencies. Developed by the National Academy of Sciences (NAS).
Emergency exposure planning guidelines (EEPG)	A health hazard factor used in computing the Dow Chemical Exposure Index. Similar to EEGLs and ERPG-2s.
Emergency response planning guidelines (ERPG)	A system of guidelines being prepared by an industry task force of which ERPG-2 is the (ERPG) maximum airborne concentration below which, it is believed, nearly all individuals could be exposed for up to 1 hour without experiencing or developing irreversible, adverse, or other serious health effects or symptoms that could impair an individual's ability to take protective action (similar to EEGLs).
Episodic event	An unplanned event of limited duration, usually associated with an accident.
Event	A specific occurrence. In this document, events are associated with specific steps in a failure sequence leading to an accident.
Hazard	A chemical or physical condition that has the potential for causing damage to people, property, the environment, etc.
Hazard zone	The zone or region where hazard impact has the potential to occur.
HTHM	High toxic hazard material.
Immediately dangerous to life and health (IDLH)	Maximum airborne contaminant concentrations from which one could escape within 30 minutes without any escape-impairing symptoms or any irreversible health effects. Developed by the National Institute for Occupational Safety and Health (NIOSH).

Impact	The ultimate potential result of a hazardous event, expressed, depending on the particular application, in terms of injuries, fatalities, environmental or property damage, business interruption, etc.
Lethal service	Service utilizing poisonous gases or liquids of such a nature that a very small amount of the gas or of the vapor of the liquid, mixed or unmixed with air, is dangerous to life when inhaled...this class includes substances of this nature that are stored under pressure, or may generate a pressure if stored in a closed vessel (ASME Boiler and Pressure Vessel Code, Section VIII, Div. I).
Likelihood	A measure of the expected frequency with which an event occurs. This may be expressed as a frequency (e.g., events per year), a probability of occurrence during a time interval (e.g., annual probability), or a conditional probability (e.g., probability of occurrence, given that a precursor event has occurred).
Maximum allowable concentration (MAC)	The maximum concentration in air of a toxic material that the facility and the public authorities having jurisdiction are willing to tolerate at a populated downwind location in the event of a major accident.
Operating instructions	A series of sequential written details describing how to operate equipment.
Operating procedures	All written information necessary to operate equipment, compiled in one document including operating instructions, process descriptions, operating limits, chemical hazards, and safety equipment requirements.
Permissible exposure limits (PELs)	Similar to TLVs (see below) but developed by the Occupational Safety and Health Administration (OSHA).
Public emergency exposure limit (PEEL)	The maximum concentration in air of a toxic material to which the public might be exposed without significant adverse impacts in the event of an accident.
Risk	The potential for realizing undesired impacts as a result of some event, facility operation, or other activity. The overall risk includes contributions from a spectrum of potential

risk elements, each having a particular likelihood and consequence associated with a specific event.

Risk assessment	The methodical evaluation and judgments involving qualitative or quantitative measures of risk.
Risk evaluation	The assessment of risk, coupled with an appraisal of the significance of the results, both overall and from individual events.
Risk management	The systematic control of hazards in order to protect employees, the general public, and the environment as well as assets and earning power.
Short-term public emergency guidance levels (SPEGLs)	For exposures whose occurrence is expected to be rare in the lifetime of any one individual, the 60-minute exposure concentration that reflects an acceptance of the statistical likelihood of a non-incapacitating reversible effect in an exposed population, while avoiding significant decrements in performance. Developed by the National Academy of Sciences (NAS).
Threshold limit value ceiling (TLV-C)	The concentration in air that should not be exceeded during any part of the working exposure. Ceiling limits may supplement other limits or stand alone. Developed by the American Conference of Governmental Industrial Hygienists (ACGIH).
Threshold limit value - short-term exposure limit (TLV-STEL)	A 15-minute, time-weighted average concentration to which workers may be exposed up to four times per day with at least 60 minutes between successive exposures with no ill effect if the TLV-TWA (see below) is not exceeded. This limit supplements the TLV-TWA where there are recognized acute effects from a substance with toxic effects that result primarily from chronic exposures. Developed by the ACGIH.
Threshold limit value - time-weighted average (TLV-TWA)	The time-weighted average concentration limit for a normal 8-hour workday and a 40-hour workweek to which nearly all workers may be repeatedly exposed, day after day, without adverse effect. Developed by the ACGIH.
Toxic hazard	In the context of these guidelines, a measure of the danger posed to living organisms by a toxic agent, determined not

	only by the toxicity of the agent itself, but also by the means by which it may be introduced into the subject organisms under prevailing conditions.
Toxicity	The quality, state, or degree to which a substance is poisonous and/or may chemically produce an injurious or deadly effect upon introduction into a living organism.
Workplace environmental exposure levels (WEELs)	Similar to TLVs, but for materials not addressed by ACGIH or OSHA. Developed by the American Industrial Hygiene Association (AIHA).

GUIDELINES FOR

Safe Storage and Handling of High Toxic Hazard Materials

1

INTRODUCTION

These Guidelines present information that is pertinent to the safe design and management of facilities for the storage and handling of high toxic hazard materials (HTHMs). The use of such materials is necessary to produce many important products used by society; organizations responsible for such activities should be diligent in assuring a high level of safety with respect to employees, neighbors, and the environment. While it is impossible to eliminate all risk, much can be done during the planning, siting, design, and operating lifetime of toxic chemical facilities to identify and avoid many chemical-related accidents, thereby reducing risks to very low levels.

To date, the chemical industry has been effective in minimizing the frequency and severity of accidental releases of hazardous materials. In addition to the existence of well-developed regulations, codes, and standards worldwide, many companies impose their own safety requirements and share them with other members of the chemical industry. These Guidelines are based on that body of knowledge, which includes good industrial practices gathered from many years of experience. They are intended to aid those people who are responsible for the safe handling and storage of HTHMs in comprehensive safety planning for new facilities, and for safety improvements in the management of existing facilities.

1.1 Scope

(a) These Guidelines cover HTHMs that are liquids, liquefied gases, and gases, and that are characterized by varying degrees of reactivity and flammability, but they exclude biological materials and lower toxicity chemicals with hazards that are primarily related to reactivity, flammability, or explosivity.

(b) They cover initial facility planning with regard to inherent safety by choice of materials, inventories, plant layout, and site selection.
(c) They deal with factors that can affect the reliability and integrity of storage vessels and how vessels should be designed, constructed, monitored, maintained, and protected against accidental release of contents, because of corrosion, overpressure, natural disasters, fires, and other potential failure modes.
(d) They cover the transfer of HTHMs to and from storage containers and the safeguards that are recommended to minimize the potential for accidents or releases and/or the quantity of chemical that may be released.
(e) The scope has been limited to storage and transfer facilities for HTHMs. While it includes rail, road, and barge transfer to permanent storage, it does not include transportation safety or offsite pipelines, which are regulated by the Department of Transportation, nor does it directly include process equipment or associated short-term storage vessels.
(f) They do not cover warehouse-type storage of HTHMs; however, the European Council of Chemical Manufacturers' Federations (CEFIC) has recently published a guidebook that addresses this topic. The CEFIC guidebook is listed in the bibliography (see Appendix A).

1.2 Applicability

(a) These Guidelines are applicable to storage facilities, including those at manufacturing chemical plants, other industrial plants, and bulk storage terminals that handle highly toxic chemicals in quantities that may result in airborne concentrations of the materials that may cause acute exposure for employees and the public in the event of an accidental release.
(b) They also apply to the permanent bulk storage of high toxic-hazard raw materials, intermediate products, and finished products.

1.3 Contents

This volume has been organized so that basic issues important in new facility design are addressed first, with subsequent chapters covering specific aspects of design and operation, which can also be readily applied to existing facilities. It is intended to extend and augment all the basic regulatory requirements for the safe storage and handling of HTHMs that must be met as minimum standards.

Chapter 2 begins with a discussion of how to identify facilities where the toxic hazards of materials are high enough to warrant the special safety measures presented in these Guidelines. Methods for the selection of a maximum credible incident for safety assessment purposes are also discussed.

Chapter 3 covers some design considerations for a new facility, including siting, design for inherent safety, plant layout, and other general factors.

Chapter 4 covers guidelines for the design of storage and piping systems, Chapter 5 presents guidelines for the design of loading and unloading systems, and Chapter 6 presents considerations for the design of associated instrumentation, sensing devices, and emergency control systems.

Chapter 7 addresses various methods for the isolation of critical system components and for the design of release containment systems, and Chapters 8, 9, and 10, which are relevant to both new and existing facilities, cover several important topics, including preventive maintenance and inspection, operating procedures and training, and emergency preparedness planning, respectively. Specific references are provided at the end of each chapter, and a bibliography listing selected relevant documents has been included as Appendix A.

2

ASSESSMENT OF POTENTIAL RISKS

The storage and handling of HTHMs involve some risks which can be reduced to very low levels by good planning, design, and management practices, as described in the following chapters. Typical facilities usually present a spectrum of risks ranging from occasional small leakages, which require prompt detection and repair, to larger releases that are extremely rare in well-managed facilities, but which may have the potential for widespread impact. A well-managed safety program should involve appropriate levels of control across the spectrum of risks.

What level of safety is acceptable is difficult to define; ultimately the choice is the responsibility of facility management and regulatory authorities having jurisdiction. Such choices have been made qualitatively in the past in the development of safety-related codes and in general industry practices. These have been improved and revised over the years as more experience was gained and as technology improved. As a result, today's chemical industry has an excellent safety record.

However, continuing vigilance is important, particularly when HTHMs are involved. Since serious chemical-related accidents are rare, it is essential to develop good techniques for identifying significant hazards and mitigating them when necessary. Hazards can be identified and evaluated using a variety of approaches such as "What If" analysis, Preliminary Hazard Analysis, Hazard and Operability Studies, and other techniques that are described in the AIChE-CCPS "Guidelines for Hazard Evaluation Procedures."

For some of the more complex, high-hazard facilities, quantitative risk assessment techniques are being used more extensively to compare relative risks, identify major risk factors, and estimate the effectiveness of mitigation options. The AIChE-CCPS is currently sponsoring development of "Guidelines for Chemical Process Quantitative Risk Assessment Procedures" which will document the techniques, data bases, and resources needed to implement a risk assessment program.

Whether the risk assessment is qualitative or quantitative, most judgments about safety levels are based on comparisons with other types of risk. For example, chemical-related risks to workers at facilities where HTHMs are stored and handled can be compared to general levels of occupational risk. Or a facility that has a good record for safe operation relative to national industry standards over the years may feel that it has a more than adequate level of safety management. When one is looking at risks to the public, again some comparisons with already existing background risks can be made to decide whether or not additional risk is significant. Chapter 9.1 of Lees[1] presents considerable information on background risks and discusses some of the complex issues involved in evaluating relative risks. Different organizations, both private and public, take different approaches to risk evaluation, and, to date, there is considerable variability among criteria for acceptability used or applied by various groups.

In qualitative terms, however, activities should be conducted by an organization in a way that protects people, property, and the environment; and ways to enhance this protection on a continuing basis should be actively sought.

When planning new facilities, there are usually more options for assuring high safety levels than for existing facilities. First, inherently safe design alternatives, as proposed by Kletz,[2] should be considered. These include

- review of raw materials and intermediates to see if less hazardous chemicals can be used,
- evaluation of storage and handling requirements to see if hazardous-materials operations can occur under less severe conditions (e.g., pressures and temperatures nearer to ambient),
- sizing of systems to reduce hazardous material inventories, and
- choice of a site and planning of the site layout based on the estimated extent of credible accidental releases of hazardous materials.

After the planning stages, final system design should be based on minimizing the likelihood of hazardous-material releases or, should a release occur, limiting the amount of release. Finally, consideration should be given to effective emergency response that can reduce the impacts of a release.

Key to safety are good management systems, which include hiring and training competent staff; having clear and well-designed operating procedures, maintenance procedures, and emergency plans; and assuring that proper procedures are followed.

Existing facilities also should be carefully reviewed to identify potential hazards when HTHMs are being stored and handled. Redesign of equipment, addition of instrumentation, and, in particular, improvements in the safety management program may be effective in reducing risks of toxic-material releases. Periodic safety reviews of such facilities are recommended, and careful scrutiny should be given to any toxic gas release incidents or "near misses," so that the source of the problem is well understood and controlled.

2.1 Safety Assessment for the Storage and Handling of High Toxic Hazard Materials

Figure 2.1 shows the major elements in an overall risk assessment program. First, in the planning stages of a project involving hazardous materials, it is necessary to identify and evaluate hazards associated with accidental releases caused by potential mechanical, human, or system failures, with consideration given to possible interactions. Some basic engineering has to be done at this stage to identify process routes, storage inventories required, and the hazardous properties of feedstocks, intermediates, products, and wastes. In selecting an appropriate site and general plant layout, qualitative evaluations of potential hazards should be made.

This preliminary hazard identification and evaluation for facilities that will be storing and handling HTHMs involves several steps (presented in question form):

- Are there HTHMs involved?
- Are any of them also reactive and/or flammable?
- What quantities are present?
- How volatile are these materials?
- What is the maximum release that is credible?
- What impacts might result?

2.1.1 Toxicological Hazards

The previous observations naturally lead to a further discussion of the difference between the "toxicity" of a substance and the "toxic hazard" it poses to the public. This is an important concept to understand because materials of high toxicity are often assumed by many to pose a severe toxic hazard, regardless of the other properties of the materials and the circumstances surrounding their release.

Imagine 1 ton of an extremely toxic, non-volatile, solid material that has been spilled in the middle of a street in a densely populated metropolis on a hot windy day. The material (e.g., sodium cyanide) is so extremely toxic that 45 pounds would be sufficient to kill 100,000 people if they somehow ingested equal portions. Alternatively, imagine an overturned tank truck on the same street that has just released 1 ton of a very common liquefied gas (e.g., anhydrous ammonia) which is considered to be moderately toxic. As it rapidly vaporizes upon release, the ton of liquid may become as much as 30,000 cubic feet or more of pure gas. Ammonia is deadly for short-term exposures at a level of 5000 ppm. If the ammonia mixed uniformly with air to reach that concentration, the lethal cloud spreading over the city would have a total volume of 6 million cubic feet.

On a strictly weight basis, although the solid material may be thousands of times more toxic than the gas, there is little chance it could poison members of the public just a short distance away because it lacks mobility. Thus, the non-volatile solid would have to be carefully handled and removed from the scene, but actually it would pose a

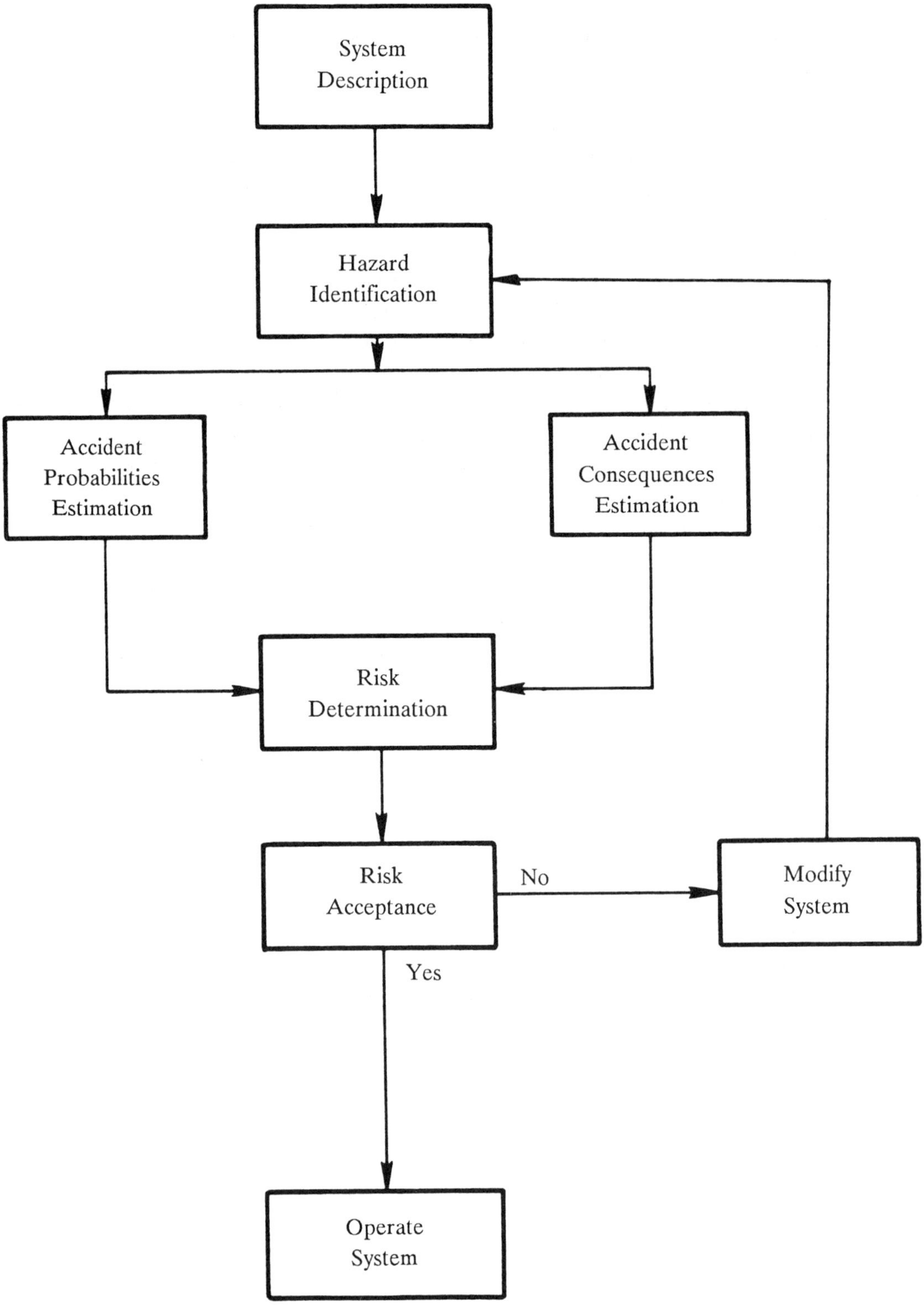

Figure 2.1. *Predictive Hazard Evaluation*

relatively low "toxic hazard" to the general public. Authorities may wish to evacuate the immediate spill area and cover the solid with plastic sheeting to prevent any dust from becoming airborne and to keep the material dry, but the risk of fatalities among the public would be low in most cases.

The situation with the lower toxicity liquefied gas would pose a far greater toxic hazard because the gas would quickly spread over downwind areas. The gas could prove fatal to people near the spill site and cause toxic effects among hundreds of others in the downwind direction.

There are several techniques for estimating the relative toxic hazard of chemicals. The Dow Chemical Company has developed a "Chemical Exposure Index" (CEI) for the relative hazard ranking of toxic materials, including factors relating to toxicity, quantity volatilized, distance to an area of concern, and physical properties. A description of the Dow methodology is presented in Appendix B. The ranking uses toxic exposure levels similar to those being developed by the Committee on Toxicology of the National Research Council[3] as Emergency Exposure Guidance Levels (EEGLs) for a 60-minute exposure. Similar values are being developed for additional materials by an industry task force and are called "Emergency Response Planning Guidelines - Level 2" (ERPG-2). These values have not yet been developed for many materials, so if a value is not available for a particular chemical, the advice of a qualified toxicologist on the inhalation toxicity of the material may be needed.

Other measures of inhalation toxicity include TLV, PEL, and WEEL which have been developed for different purposes. A key problem of using these limits in the course of emergency planning is that they are primarily intended for use in the occupational environment where presumably healthy workers are exposed to concentrations near these limits day after day throughout their careers. This consideration and the desire to prevent health effects associated with chronic exposures mean that these values are often (not always) much lower than what they have to be to protect the public from exposures associated with rare or infrequent spills of hazardous materials. Consequently, use of a TLV, PEL, or WEEL value, although decidedly safe in the vast majority of cases, could conceivably result in a major overprediction of downwind hazard zones in many cases. Exceptions involve those materials, such as chlorine, for which limits have been based on irritant rather than toxic effects.

NIOSH's IDLH limits are considerably higher, are defined for an exposure duration that is closer to what would be expected in an actual short-term spill emergency, and are closer to the borderline between levels that are barely tolerable and those that may cause significant injury. The problem is that "barely tolerable" contaminant concentrations may actually cause considerable irritation or other distress, possibly to the point of prompting large numbers of people to seek medical assistance. Also, since NIOSH is again assuming that healthy workers are being exposed, IDLH concentrations may actually cause significant injury to young children, the elderly, or individuals with pre-existing health problems. For this reason, some chemical facilities are known to be considering use of the IDLH value divided by a safety factor for emergency planning purposes (or the TLV-TWA or TLV-C if either has a higher value than this result).

The NAS/NRC SPEGLs and AIHA ERPG-2 values are possibly the best choice among currently available guidelines, given the objectives of their respective developers. Unfortunately, only a few hazardous materials have been addressed to date.

To use the DOW CEI in the absence of toxic limits equivalent to their EEPG values (or equivalent EEGL or ERPG-2 values), an assumption may be needed. The equivalent value should lie between the IDLH values and the occupational safety levels. CEI values in excess of about 300 indicate a HTHM facility. Because of the uncertainties involved in such preliminary ranking systems, rankings just below a category boundary should be reviewed as possibly being in the higher risk category.

Another screening system is given in the Environmental Protection Agency's (EPA) interim rulemaking 40 CFR 300 in response to the Superfund Amendments and Reauthorization Act of 1986 (SARA, P.L. 99-499). Under Supplementary Information to these interim rules, a method of ranking is given as:

$$\text{Index} = \frac{\text{Level of Concern}}{V}$$

where the IDLH level (in milligrams per liter) is used as a surrogate measure of the level of concern, and V is the extent to which the chemical can become airborne. For gases, $V = 1$. The regulation gives methods for estimating V, a vapor fraction, for evaporating and boiling liquids. The lower the value of the index, the higher the toxic hazard. The EPA gives ranges of the index that they used to find threshhold planning quantities for their rules. For example, for values of the index that are less than unity, the EPA uses threshhold quantities of 500 lb or less.[4]

The SARA Index is a rougher screening tool than the Dow CEI (Appendix B) and does not actually represent the relative toxic hazard for different quantities stored. Potential hazard zones increase with quantity released in less than a linear fashion, so it is difficult to adapt the SARA Index to reflect the toxic hazard. Similar methodology for relative ranking of toxic hazards among a number of installations may be developed by utilizing equivalent, or better, measures of material toxicity and volatility.

2.1.2 *Design Basis Incident (DBI)*

For the purposes of siting, facility layout, and emergency planning, it is necessary to have an idea of the types of accidental releases of a HTHM that might occur over the lifetime of a facility. The DBI is usually selected as the release scenario with the largest impact that might credibly occur during facility operations.

These Guidelines address storage and handling facilities for HTHMs, so potential release scenarios might include anything from storage tank failure to major transfer line failure, to release through a relief valve to breakage of a small fitting off a line, or leakage from a pump or valve packing. But, what is credible?

HTHM storage tanks that are designed to accommodate the hazards of the specific materials involved, and have careful construction, testing, operation and maintenance, rarely fail in a catastrophic manner, unless exposed to extreme external hazards such as fire or earthquakes. Tank design (including design for extreme winds, earthquakes, snow loads, etc.) can be used to make such sources of failure virtually impossible. In such a case, a total tank failure might well be considered "non-credible." Of course, for some older facilities where documentation on the design and operational history of older tanks is missing, where flammables may be stored close to toxic materials, where the stored toxic material is also flammable or reactive, or where a large number of tanks are closely clustered, it may be prudent to consider a tank failure as the DBI. A more realistic DBI for most modern, well-run facilities is usually a failure of a major transfer line.

In analyzing potential release scenarios in a realistic manner, it is useful to identify causes of failure, release locations, and quantity of toxic release. In piping systems, the inventory might be the contents between shut-off valves, including flow during the time the valves are being closed. The valves might be automatically actuated, or might require some time for an operator to take appropriate action. Usually a conservative assumption is made that a full-bore failure would occur, releasing inventory between shut-off valves plus an additional volume corresponding to the maximum transfer rate during the total time it would take to close the valves. This time could range from less than a minute for automatic shutdown systems to 10 minutes or more, if manual shutdown were required.

If a transfer line is connected to a storage tank, nozzle failure at a penetration below the liquid level could drain the tank to that level if no positive shut-off valve were fitted. Failures of pipelines connected to dip pipes may have the same effects if the tank is under pressure or without a siphon break.

In all these examples, great care should be taken in analyzing the design basis accident scenarios to make sure they are realistic. A full-bore failure of a large-diameter pipe might also be deemed "non-credible" if the line can be shown to have been designed, constructed, and protected against all major sources of potential failure including metallurgical defects, third party damage, fires, and explosions. For such high-integrity systems, a DBI might be the failure of a more vulnerable smaller line or component.

For purposes of initial siting, it is important to select a DBI that is conservative, and this is why full-bore piping failures are often selected, even though such failures are also quite rare.

After a design is completed, a more accurate assessment of potential release scenarios can be conducted in cases where this may be warranted, even up to levels of detail associated with quantitative risk assessment. Modifications in design or operation can be made where necessary to reduce either the likelihood or potential consequences of credible accidents.

2.2 Toxic Chemical Hazard-Zone Estimation

The extent of hazards to human life and health associated with a particular toxic chemical release can be predicted using available mathematical modelling techniques described in the "Guidelines for the Use of Vapor Cloud Dispersion Models," published by the AIChE-CCPS.[5]

All these techniques have several steps in common. First, the kind of release must be described by assessing the chemistry and physical state of the material released, the release rate, and the duration of the release. For example, if acetone cyanohydrin is released, its decomposition to acetone and hydrogen cyanide causes a hydrogen cyanide toxic hazard. The release could be a vapor jet, a jet with entrained aerosols, or a spill of a volatile or boiling liquid into an impounding area or on the ground. The estimates of release rates have to be based on actual conditions such as choked flow or two-phase flow within pipelines under release conditions. When a liquefied gas, like ammonia or chlorine, is released to atmosphere, a major portion of it vaporizes rapidly ("flashes") and much of the residual liquid portion may be entrained as aerosols. Likewise very fine powdered solids may also become airborne if released under high-wind conditions. Such airborne aerosols and particulates then may disperse downwind in a fashion similar to a dense gas.

If mixtures are released, a higher fraction of the more volatile components will flash, leaving the residual liquid rich in the heavier components. When a liquid mixture is spilled, the more volatile components will vaporize preferentially. These are discussed in more detail in the AIChE-CCPS's "Guidelines for Use of Vapor Cloud Dispersion Models."

After a release, the toxic material is then carried downwind while it mixes and spreads until diluted to safe limits. Atmospheric conditions influence impact areas. Dispersion is slowest under conditions of stable weather and low-wind speeds. The density of the dispersing vapor and the topography are also important. Dense gases will mix by gravity spreading, while buoyant gases may tend to rise, thus reducing ground-level risks. Very short duration spills may be modelled as "instantaneous releases" that move downwind as a puff while dispersing. Longer duration "continuous spills" form plumes that may reach steady-state downwind concentration levels until the spill source is shut off. Many actual spills lie between these two limiting behaviors and both spill types should be modelled.

Dispersion calculations are subject to considerable uncertainty, but they are useful in considering potential downwind hazard distances, and hazardous impact areas for purposes of safety management planning. These models can also be used to calculate the maximum quantity of release that would disperse to safe levels before reaching a location of concern. This may be useful, for example, in screening alternative sites, for determining the maximum inventory of hazardous materials that may be kept on-site without posing risks to populated areas, for the sizing of individual storage tanks, for the placement of major inventories within a site, and for the location of isolation valves.

If the chemicals involved are also reactive or flammable, these additional effects should also be considered. Flammable limits of materials are given in the National Fire Protection Association (NFPA) "Fire Protection Guide on Hazardous Materials" (containing NFPA 325M/49/491M/ 704)[6] and in other references. For highly toxic flammable vapors, the toxic hazard zone is generally larger than the flammable zone. However, if the cloud is ignited, the fire tends to burn back to the source. The combustion plume often may rise to limit further downwind travel of toxic vapor beyond the ignition point. However, some residual toxic vapors may persist downwind. Combustion or decomposition upon heating of some materials may also generate highly toxic vapors (e.g., hydrogen sulfide burns to sulfur dioxide, and chlorosulfonic acid decomposes into sulfuric acid and hydrogen chloride).

Reactivity information is available from a number of sources including the NFPA Guide (cited above).[6-8] When a toxic chemical is highly reactive in the presence of certain contaminants, this becomes an important design consideration. Also, chemicals that can undergo runaway reactions, decomposition, or self-polymerization with resulting temperature or pressure increases may be subject to problems if storage and handling systems are subject to upset, contamination, or heating by an external fire. Where flammable materials are stored in proximity to highly toxic materials, close scrutiny is needed to assure that external fires do not threaten the integrity of toxic material containment.

When hazard zones are estimated for credible releases, the potential impacts within these zones should be evaluated for use in safety planning. Are operators, employees in adjacent plant areas, or those in nearby office structures adequately protected? Are ventilation systems within potential impact areas designed to prevent intrusion of toxic vapor? Have the people who live near the facility been properly protected against the impacts of credible accidents? For the highly toxic chemicals, if stored and handled in quantities of concern, consideration of such potential problems should be considered carefully in site selection. Any residual risks should be evaluated in emergency response planning to minimize potential impacts.

References

1. Lees, Frank P., "Loss Prevention in the Process Industries," Vol. I, Butterworth, Boston and London, 1980.
2. Kletz, T., "Cheaper, Safer Plants or Wealth and Safety at Work," The Institution of Chemical Engineers, Rugby, England, 1984.
3. "Emergency and Continuous Exposure Guidance Levels for Selected Airborne Contaminants," Vols. 1-6, National Research Council, 1984-85.
4. Federal Register, Vol. 51, No. 221, p. 41577, 17 November 1986.
5. "Guidelines for Use of Vapor Cloud Dispersion Models," AIChE-CCPS, New York, 1987.
6. "Fire Protection Guide on Hazardous Materials," 8th ed., containing "Manual of Hazardous Chemical Reactions," NFPA 491M-1975, National Fire Protection Association, Batterymarch Park, Quincy, MA 02269, 1984.

7. Bretherick, L. "Handbook of Reactive Chemical Hazards," 3rd ed., Butterworth, Boston and London, 1985.
8. Hatayama, H.K., et al., "A Method for Determining the Compatibility of Hazardous Wastes," EPA Report No. EPA-600/2-80-076, Municipal Environmental Research Laboratory, U.S. Environmental Protection Agency, Cincinnati, Ohio, April 1980. Available as publication PB80-221005 from the National Technical Information Service, Springfield, Virginia 22161.

3

DESIGN CONSIDERATIONS FOR A NEW FACILITY

This chapter covers the design considerations for a new facility for the storage and handling of HTHMs, which may be volatile liquids, pressurized liquefied gases, refrigerated liquefied gases, or gases. In planning for such a new facility, important safety options are available that may not be feasible alternatives for existing facilities. While these Guidelines focus on the storage and handling of HTHMs, planners initially should look at safety in the broadest context, relative to process and site selection since trade-offs may be involved. For example, siting away from populated areas may yield low risks for fixed facilities, but may simultaneously increase transportation risks. Also, large separation distances between storage tanks may reduce some risks, but increase others associated with longer piping runs.

Many of the codes and standards used for siting and plant layout have evolved from fire codes and regulations based upon concerns about fires and explosions. The National Fire Protection Association (NFPA) standards, which are widely used, but based on fire and not toxicity, state that their separation distances and other safety requirements are intended as minimum acceptable standards. For example, many of the separation distances are based on concern for off-site fire spread and may not assure safety without backup of firewater systems. In general, their standards for relief valve sizing consider only fire exposure, and not other relief requirements, such as runaway reactions. Many of these standards have been adopted by the Occupational Safety and Health Administration (OSHA) and as state and municipal regulations and, therefore, cognizance of their existence is necessary. Regulations for HTHMs are less developed than those for much more common flammable materials, and often they do not address the potential for large releases of toxic materials. When gas releases of toxic chemicals are involved, NFPA requirements usually do not provide adequate protection for the public, since toxic vapor clouds may sometimes impact greater areas than fires or explosions. There are no widely published codes and standards that recommend minimum separation distances from storage tanks to property lines and important off-site occupancies, based solely on toxicity.

3.1 Inherently Safer Design

In the past, plant designers may have assumed that adequate safety was assured by compliance with applicable codes and standards and with generally accepted industry practices. But today, in the aftermath of several recent tragic events*, most owners, operators, and regulators of hazardous material facilities are increasingly aware of the importance of closely scrutinizing hardware and operations, whether existing or planned, to find potential weaknesses and to reduce the risks to employees and neighbors.

The use of inherently safer practices, as discussed by Kletz,[1] is an approach to the design of safer chemical plants and storage facilities that is currently receiving much attention and application. The approach is simple in concept:

> First, explore alternatives that will allow the same product to be made using an inherently safer technology. This includes using materials that are less hazardous, and then processing, transporting, and/or storing these materials under potentially less hazardous conditions, and incorporating robust and proven equipment, systems, and designs. Sometimes research and development efforts may be needed. Such an alternative, when identified, is obviously preferable unless there are extenuating circumstances.
>
> Second, assess the safety of the baseline technology and judge its acceptability. This evaluation not only allows the designer to decide if the design is "safe," but also allows design alternatives to be compared so that improvements in safety can be recognized and achieved.

Once a process design has been chosen, the focus shifts to limiting the amount of a hazardous material release and reducing the probability of that release. When hazardous materials must be handled or stored, various techniques[2,3] (e.g., fault-tree analysis) can be used for evaluating and reducing the likelihood of a release. The safety of that facility will be further enhanced by reducing the inventory of hazardous material, and by storing the hazardous material under less severe conditions (i.e., usually at lower pressure, lower temperature, and, in some situations, lower concentrations).

Finally, systems and equipment can be employed to minimize the consequences of any releases that might occur. When designing any facility, all of the approaches to safety should be appropriately utilized. The sequence of use may be important, because it is generally best to eliminate or minimize a potential hazard by inherently safe design if that is feasible. After this is done, one can explore means to limit the potential for, and the size of, releases of HTHMs and, finally, to provide prudent release detection, mitigation, and emergency response measures.

---

* For example, fire and BLEVEs in Mexico City, toxic gas release in Bhopal, and waterway contamination in Basel.

It is obvious that the most efficient way to utilize these concepts for a safer chemical plant is to incorporate them into the design at the earliest possible stage -- in some cases in the initial research and development phase. Researchers who are developing new process technology should be familiar with the concepts of inherent safety in chemical plants and storage facilities. They could pursue equivalent technologies using less hazardous materials or less hazardous storage conditions. It should also be noted that, although these safety concepts are most effective when incorporated during the initial design, it is not implied that effective safety improvements cannot be made after the facility has been constructed and is operating. These concepts can also be applied to existing facilities, and the feasibility of retrofitting an operating facility for enhanced safety should be considered on a case-by-case basis.

Considerations for inherently safer facilities are summarized below:

- Use a less hazardous material, if feasible.
- Reduce inventories of hazardous materials to the extent feasible by optimizing the quantity in any container and the number of containers, controlling the quantity of chemical between emergency shutoff valves, optimizing the quantity used or shipped, and considering the use of storage alternatives, such as on-site continuous feedstock production and the use of close-coupled reactors to eliminate the need for intermediate storage.
- Utilize less severe process conditions.
- Use fail-safe and/or redundant safeguards for safety systems, emergency shutoff valves, and emergency block valves.
- Substitute passive protection systems for active ones, if practical (e.g., prevention of frost heaves under a refrigerated tank using an air space under an elevated pile cap rather than heaters in a grade-level foundation, or tank separation rather than water deluge for protection from a potential external fire exposure).
- Use corrosion-proof materials of construction to lessen the need for periodic inspection, or the possibility that an upset will cause accelerated corrosion.
- Minimize piping carrying HTHMs outside unit limits, with such piping, as required, protected from vehicular traffic and other perils.
- Provide periodic safety reviews of design and operations.[2] Reviews of modifications are particularly critical.

3.2 Basic Siting Practices

The careful selection of a site for a storage facility can reduce the potential off-site impacts of toxic material releases. Some good practices currently in use are listed below:

(a) Site the storage facility such that sufficient controlled-use "green space" exists to provide a buffer zone based on the DBI between the release location and the nearest population. For long-term effectiveness, the "green space" should be under the control of the plant's owner.

(b) Favor storage sites with the lowest potential population exposure and ascertain that transportation risks are acceptable. The extent of toxic hazards for credible releases should be evaluated and considered when selecting a site and planning the location for major inventories of toxic chemicals. The census statistics do not necessarily reflect the transient population conditions, such as schools, commercial and industrial areas, and hospitals. Emergency evacuation plans and warning systems must be carefully considered when siting a major toxic chemical facility.

(c) Site the storage facility to provide adequate separation from other facilities in order to avoid "knock-on" or secondary damage caused by blast waves, debris, missiles, or fire radiation from other episodic events.

(d) Site the storage facility to support the heavy loads that storage tanks usually entail, and use good civil engineering practices. Provide adequate flexibility in piping systems without sacrificing strength. Careful consideration should be given to earthquake potential at all sites. Except in active fault areas, adequate design can usually protect piping and storage systems against rupture; clearly, siting a storage facility for HTHMs in an active fault area should be avoided.

(e) Site or grade the facility such that terrain contours reduce the likelihood of liquids and heavy vapors flowing toward populated areas.

(f) Site the facility so that the probability of damage from external sources is minimized. Adequate clearance or crash barriers should be provided to protect against vehicular impact. Do not locate the facility within or close to landing and takeoff patterns of an airport, since most aircraft accidents take place during these maneuvers. Make sure designs include consideration of windstorms, floods, landslides, subsidence, and other potential external threats, such as vandalism, terrorism, and sabotage. Provide a well-thought-out site security program/system.

(g) Site the facility such that adequate roads for routine and emergency situations are available or can be built. An adequate number of unblockable roads should be available for the movement of employees, members of the public, ambulances, and emergency service vehicles travelling both to and from the facility. [A blockable road is considered to be a narrow road (less than 2 full lanes) or a road that narrows for a bridge crossing or an underpass or has a railroad track crossing, especially one used for switching cars.] Also, make sure there are at least two such access roads in case one is rendered impassable by the emergency.

(h) Site the facility so that, in the event of a liquid release, water supplies will not be contaminated.

(i) Choose a site where adequate reliable emergency utilities (power, water, etc.) can be provided.

(j) Site the storage facility such that nearby populations, both off-site and on-site, are up-wind of the prevailing wind. Since prevailing wind speed and direction are not totally reliable for safety planning, this approach should be used only as a supplementary factor.

3.3 Plant Layout

The potential severity of an accidental gas release from a toxic chemical storage area may be greater than that from the process area or from the loading/unloading area, because of the larger quantity of material in storage. Typically, two orders of magnitude exist between the quantity of materials in storage and those in process for a large plant. On the other hand, the frequency of storage accidents is quite low, compared with loading and unloading accidents and process accidents. This poses the design problem of keeping the storage tanks for HTHMs sufficiently far from property lines, while maintaining adequate clearance distances from loading/unloading and process areas and, simultaneously, keeping the loading/unloading and process operations a safe distance from populated areas.

There must be adequate separation distances between storage facilities and other hazardous operations and sensitive areas on the site, such as:

- loading/unloading areas,
- process operations and control center,
- other tank farms,
- utilities,
- services, and
- administration areas.

Separation distances for gaseous releases based on toxicity are currently not defined by code and therefore should be evaluated individually based upon specific site factors, that is, inventory, release rate, topography, meteorology, and emergency planning. One can then use dispersion studies to evaluate the hazard distances.

The following additional guidelines should be considered for planning the layout of a HTHM storage area:

(a) High-integrity storage tank design is essential for a HTHM. Consideration should be given to overdesign over normal operating conditions, or the use of double integrity storage tanks, for example, with a concrete outer tank or berm.

(b) If the liquid storage tank for a HTHM is diked, it is good practice that the same dike does not contain inventories of other materials that are flammable, reactive, or non-compatible with the HTHM. If located next to a diked area containing flammable and/or reactive chemicals, the toxic storage may have to be provided with fixed fire water protection for cooling. Factors to be considered include drainage in the diked areas, the response time, the adequacy of the site fire brigade or public fire department that would respond, and the flammability and reactivity of the nearby materials. The impounding area should be designed to minimize the vaporization rate of volatile HTHM. Installed means of evaporation suppression should be considered (e.g., foam).

(c) A HTHM should not be stored in proximity to explosive/flammable/ reactive materials that might impact the storage tank. Diking for materials that are flammable or reactive should be sloped or drained to carry spilled material to a separate impoundment tank or area and away from any nearby HTHM tank. For HTHMs the impoundment system should be designed to minimize the surface area and move material away from other hazard areas.

(d) The concept and design of a designated dump tank should be investigated at the earliest stages of site layout. Depending on the potential hazard and the time required to empty the HTHM storage tank, it may be desirable to keep some storage vessels empty, or at low levels, for such emergency transfers. The possible need to empty a leaking tank is one important reason for providing a submerged pump or other transfer method.

(e) The momentum of the tank contents might overflow a traditionally designed dike wall if the tank were to burst. A dike wall of about one-half tank height is required for containment of this scenario. Full-height, close-in dikes or berms have been provided for some cryogenic ammonia (and liquefied natural gas) storage tanks. The dike must be sized to contain at least the volume of the largest tank within the dike.

References

1. Kletz, T. A., "Cheaper, Safer Plants or Wealth and Safety at Work," The Institution of Chemical Engineers, Rugby, England, 1984.
2. "Guidelines for Hazard Evaluation Procedures," AIChE-CCPS, New York, 1985.
3. Lees, F. P., "Loss Prevention in the Process Industries," Vols. I and II, Butterworth, Boston and London, 1980.

4

DESIGN OF STORAGE AND PIPING SYSTEMS

This chapter covers the design of the storage vessels and the piping and appurtenances for HTHM storage containment systems. Low-pressure storage vessels should meet, as a minimum, the American Petroleum Institute (API) 620 Standard, and pressure storage vessels should meet the requirements of the American Society of Mechanical Engineers (ASME) Boiler and Pressure Vessel Code, Section VIII. Piping systems for storage tanks for HTHMs, as a minimum, should meet the requirements of the ASME Code for Pressure Piping B31.3 - Chemical Plant and Petroleum Refinery Piping. Code B31.3 also covers pressure piping systems and systems designed for pressures under 15 psig, if the fluid is flammable, toxic, or damaging to human tissue. For the purposes of this chapter, Code B31.3 will be referred to as the Pressure Piping Code, the Piping Code, or simply as the Code when confusion will not arise as to whether the Pressure Vessel Code or the Pressure Piping Code is being discussed. A partial listing of other codes and standards having helpful safety design information is presented in Tables 4.1 through 4.5 at the end of the chapter. The date of the most recent edition at the time of this publication is indicated.

4.1 Vessel Design

Low-pressure storage vessels for a HTHM should be designed according to API Standard 620, "Recommended Rules for the Design and Construction of Large Welded, Low-Pressure Storage Tanks." This standard covers tanks designed for all pressures under 15 psig. It allows low-pressure tanks to be constructed in general accordance with the unfired pressure vessel codes, except that higher allowable design stress values are normally used.

Appendix R of API Standard 620 is used for the design of tanks for refrigerated products stored at temperatures ranging from 40° to minus 60° F. Appendix Q of the same Standard covers the design of tanks for the storage of liquefied hydrocarbon gases at temperatures not lower than minus 270° F. There are no specific requirements in this code based on HTHMs.

Pressure storage vessels (over 15 and under 3000 psig) for a HTHM, as covered in this document, fall within the scope of the ASME Boiler and Pressure Vessel Code's Section VIII "Pressure Vessels Division I," and thus should be designated as lethal service. Note that the Code states: by lethal substances are meant poisonous gases or liquids of such a nature that a very small amount of the gas or of the vapor of the liquid mixed or unmixed with air is dangerous to life when inhaled. ... this class includes substances of this nature which are stored under pressure or may generate a pressure if stored in a closed vessel." This is similar, but not exactly the same definition as that for "Category M Fluid Service" of the ASME Pressure Piping Code. As with the Piping Code, the owner has overall responsibility for compliance with the Vessel Code and for establishing the requirements for design, including the determination of "lethal service." Vessels should be repaired by qualified welders (Code certified if jurisdictionally required) and the vessel plans and Code certificates should be kept up-to-date.

For substances handled in small amounts, vessel design according to Code can result in relatively thin walls for low design pressures. For extremely toxic materials, such as arsine, it is often desirable to use a heavier wall design. This can be achieved easily with the use of compatible DOT cylinders. Their use provides a heavy-walled vessel with a higher than necessary pressure rating at a reasonable cost. Refrigerated storage at atmospheric pressure may be the preferred mode of storage for large quantities of highly toxic liquefied gases, after all trade-offs of high pressure, ambient temperature versus low pressure, and low temperature are investigated.

Another document that addresses chemical storage is API 650, "Welded Steel Tanks for Oil Storage." This standard has limited applicability to HTHM storage, since it prohibits refrigerated service and limits pressures to 2.5 psig and only if designed to Appendix F. Most API 650 tanks have an "internal working pressure approaching atmospheric pressures," and hence, their pressure relieving devices must vent directly to atmosphere. Its safety factors and welding controls are less stringent than those inherent in or required by API 620. It is also this standard that covers floating roof tanks and tanks with internal floaters, which are not suitable for HTHMs because of the difficulty in controlling and containing the leakage of vapors around the seals of the floating sections.

4.2 Additional Guidelines for Low-Pressure Tanks (API 620)

The Guidelines in this section are engineering safeguard practices that should be considered, and they may be applied in design, construction, and fabrication of low-pressure tanks for a HTHM.

4.2.1 Design Guidelines (API 620)

Horizontal and vertical cylindrical tanks are most often used to store highly toxic liquids at atmospheric pressure. Horizontal, vertical, and spherical tanks can be used for refrigerated liquefied gases that are stored at atmospheric pressure. Generally, because of the economics, horizontal tanks are used for small-quantity storage, spheres for medium quantities, and vertical tanks for the largest quantities.

Within the context of API 620, the design pressure of tanks for atmospheric and low-pressure storage at ambient temperature should not be less than 100% of the vapor pressure of the material at the maximum design temperature. The maximum design metal temperature is the maximum temperature to which the tank will be subjected, taking into consideration the maximum temperature of the material entering the tank and the maximum ambient temperature (include solar radiation effects if the tanks have a dark absorptive surface) of the area where the tank is located. This Code can only be applied when the temperature of the tank or entering materials is 200° F or less.

For flammable materials that are stored at conditions above their flash points, vertical vessels covered by API 620 or 650 are sometimes designed so that the roof seam is weaker than the rest of the tank. This permits the vent or pressure relief system to be sized for normal operating loads (e.g., breathing, or pump-in/pump-out); for pressure relief for fire exposure, the roof separates from the tank. For this reason, the use of a weak roof-seam is not normally acceptable for HTHM tanks.

Small horizontal and vertical tanks are best designed and stamped for 15 psig to provide for the maximum safety and latitude in the use of the tank. Since HTHM discharges simply cannot be released to the atmosphere, but must be contained and treated, it is often more cost effective to design the tank for a higher pressure and reduce the likelihood of discharges.

The minimum design temperature should be the lowest temperature to which the tank will be subjected, taking into consideration the minimum temperature of materials entering the tank, the minimum temperature to which the material may be auto-refrigerated or mechanically refrigerated, and the minimum temperature of the area where the tank is located. API 620 provides for installations in areas where the lowest recorded one-day mean temperature is minus 50° F.

API 620 covers the design of both single-wall and double-wall metal refrigerated tanks. In event of minor leakage from the inner shell, the outer shell will contain the refrigerated liquid. However, API 620 does not require the outer shell to be designed for the temperature of the lading, nor for the hydrostatic head that it would sustain if the inner tank containment failed in a major way. Sometimes the outer tank is constructed of prestressed or reinforced concrete and offers some protection of the inner tank from external damage. These designs have been developed for use in some applications involving storage of very large quantities of refrigerated flammables (e.g., liquefied natural gas or refrigerated propane).[1] There are few applications involving a HTHM where their use is suitable. One exception might be large-scale storage of refrigerated ammonia; however, a conventional tank design with a high close-in dike is a more common alternative.

There are several variations of this type of design. The Chemical Industry Association Code of Practice for the Large-Scale Storage of Fully Refrigerated Anhydrous Ammonia in the United Kingdom recommends a design where the inner shell is open topped and only the refrigerated vapors are in the space between the two tanks. The tank is insulated only on the outside. This design provides a passive limitation on the external pressure that can be exerted on the inner tank. For a full double-tank design, the pressure in the annulus must be carefully controlled to protect the inner tank. However, if an inert gas is used in the annular space, it also can be monitored to check for inner tank leakage.

While either rupture disks or relief valves are allowed on storage tanks by Code, rupture disks by themselves should not be used on tanks for the storage of HTHMs since rupture disks do not close after opening and may lead to continuing release of toxics to the atmosphere.

The API 620 Code requires a combined pneumatic-hydrotest at 125% of design tank loadings. In tanks designed for low-density liquid, the upper portion is not fully tested. Thus consideration should be given to testing and designing all HTHM tanks for hydrotesting at the maximum specified design liquid level. This may require that the lower shell thickness be increased to take the full head of water, and that the foundation be upgraded so that it can support the tank when full of water. However, the benefits of increased assurance against tank failure are justified for large refrigerated tanks. Testingin this manner not only tests the containment capability of the tank, it also provides an overload test for the tank and the foundation similar to the overload test for pressure vessels. In addition, it provides some degree of mechanical stress relief for welding-induced stresses and consequently decreases the chances of brittle fracture. Acoustic emission tests can be done conveniently during hydrotesting to provide a baseline for future testing.

In addition to radiography required by API 620, it is recommended that all T-welds and any other weld subject to hoop stress be 100% radiographed, and that consideration be given to 100% radiography of all accessible welds. All other welds should be subjected to magnetic-particle, or liquid-penetrant, or ultrasonic examination. Magnetic-particle inspection can provide a baseline against which to assess inspections made in the future. In all phases, quality control and materials control should be maintained and documented.

For general information regarding foundations for low-pressure tanks in large sizes, the reader is referred to Appendix C of API 620, "Suggested Good Practice Regarding Foundations." Proper preparation of the subgrade and grade is extremely important for tanks that are to rest directly on grade. Bottom corrosion and uneven settlement with tank distortion can result from poorly drained, non-uniform, and inadequate foundations. Foundations and subgrades should be designed to safely carry the weight of the tank and its contents when filled with water to the highest level required for the hydrostatic test.

Precautions should be taken against ground freeze under refrigerated tanks. Such freezing can cause the ground to heave and cause damage to the foundation and/or the tank itself. Free circulation of air beneath the tank is a method for passive protection from ground freezing. This is particularly useful when the tank is supported by piling because of poor soil load-bearing characteristics, and the tank can be kept off the ground using pile caps. Alternatively, electrical heating elements or heating coils containing heated low-freezing-point liquid should be placed throughout the foundation below the base insulation. Thermocouples should also be embedded in the foundation to monitor its temperature. The heating elements should be protected from high ground water level or flood. Provisions should be made for withdrawing and repairing any nonfunctioning elements.

4.2.2 Construction Guidelines (API 620)

Stress relieving when required by API 620 should be done in the shop whenever possible. Subsequent welding operations may require local stress relief in accordance with API 620. A site-erected tank must also be stress relieved in accordance with API 620. The reader is referred to Appendix H, "Recommended Practice for Use of Preheat, Postheat, and Stress Relief" and Appendix I, "Suggested Practice on Peening." Both are found in API 620.

4.2.3 Materials Guidelines (API 620)

The proper choice and testing of materials used for refrigerated tanks are both of vital importance. Steels lose their ductility at low temperature and become subject to brittle fracture and catastrophic failure. Appendices Q and R of API 620 offer specific requirements for the metals to be used, along with the Charpy V-notch values of metals suitable for the primary components of a refrigerated storage tank. When impact tests are required for primary components, they must also be made for each welding procedure. Charpy V-notch specimens must be taken from the weld metal and from the heat-affected zone of the welding qualification test plates or duplicate test plates. When impact tests are not required for the material to be welded, the manufacturer must provide evidence that the weld metal and the heat-affected zone will produce suitable Charpy V-notch impact values. Impact tests for welding procedure qualification are suggested whenever heavy-pass welding is to be used on primary components that have not been impact tested. Primary components are those whose failure would result in leakage of the liquid being stored, those exposed to the refrigerated temperature whether or not in contact with the refrigerated liquid, and those subject to thermal shock.

Materials of construction must be carefully chosen for compatibility with corrosive chemicals and external exposure. The corrosion allowance provided in the design should be adequate for the projected life of the storage tank. Welding reduces the corrosion resistance of many alloys, leading to localized attack at the welds or heat-affected zones in certain environments. This may be avoided by use of special grades of certain alloys and weld materials, or by an annealing heat treatment of an entire vessel after construction.

Besides straight-forward chemical attack, certain alloy-environment combinations suffer stress corrosion cracking (SCC), which may result in catastrophic failure. Examples of such combinations are cold-formed brass in ammonia, austenitic stainless steels in organic chlorides, Alloy 400 in hydrofluosilicic acid, and steel in hot sodium hydroxide.[2] All stresses contribute to SCC, including fabrication stresses (from welding), thermal stresses, and mechanical stresses. Fabrication stresses may be reduced by a post-fabrication (post-repair) stress relief heat treatment, but other stresses may still be adequate to cause SCC.

External corrosion can also endanger the tank. Austenitic stainless steel tanks subjected to atmospheric chlorides (such as from ocean spray) and warm temperatures are vulnerable to SCC. Corrosion may also occur under insulation, particularly if the tank is operating at temperatures where condensation is likely, or if the weather barrier is not properly maintained. Corrosion under insulation is particularly insidious, because the damage is hidden and may progress unseen for long periods of time. Special attention should be paid to the areas around supports and nozzles. Where acid gases are likely to be in the air, special external coatings may be necessary. The reader is referred to Appendix G of API 620, "Suggested Good Practice Regarding Corrosion Allowance," for additional information.

By the use of fracture mechanics, it is possible to determine whether a vessel is operating in the low-risk, leak-before-fracture (LBF) range or in the high-risk fracture-before-leak (FBL) range. Fracture mechanics is based upon the concept that, when a body containing a crack or crack-like defect is loaded in such a way that the stress intensity factor at the crack tip reaches a critical value, unstable crack growth (or propagation) will occur. Simplified formulas have been developed that allow the rapid determination of whether a vessel is being operated in the FBL mode and consequently could fail catastrophically without any advance warning, or in the LBF mode and consequently would give warning of an impending failure. Analysis must be done over the entire possible vapor pressure range of the compound stored. For the FBL mode vessel, changes in the operating conditions may vary the mode to LBF. If this is not possible, a more stringent inspection program and more frequent acoustic emission and other testing are necessary. Anyone who has critical vessels storing HTHMs is urged to investigate the subjects of fracture mechanics and failure mode diagrams and their applicability to those vessels.[3]

4.3 Additional Guidelines for ASME Pressure Code Tanks

The guidelines in this section are engineering safeguard practices that should be considered. They may be applied in design, construction, and fabrication of ASME pressure tanks for HTHMs. The following guidelines are in addition to the requirements of the ASME Code.

Vessels should be repaired by qualified welders (Code certified if jurisdictionally required), and the vessel plans and Code certificates should be kept up-to-date and valid.

4.3.1 Design Guidelines (ASME)

Pressure vessels used for the storage of HTHMs, which could be exempted by ASME because of size or high pressure, should still be designed and constructed in accordance with the requirements of the ASME Code.

Some chemicals become more corrosive if they are contaminated with water, air, or other materials (e.g., dry chlorine is customarily handled in carbon steel, but it becomes very corrosive if small amounts of water are present). The possibilities of such contamination and corrosion should be carefully considered during the design phase, and if they cannot be reliably prevented, it is possible to increase the corrosion allowance or use a more resistant material of construction.

The Code requires that the corrosion allowance be adequate to compensate for the more or less uniform corrosion that will take place during the expected life of the vessel and not weaken the vessel below the design strength. In addition to uniform attack, the corrosion allowance should factor in pitting, erosion, abrasion, oxidation, and any other deterioration that the tank may undergo from the inside or the outside. Since stress (either thermal or mechanical) increases rates of corrosion, particularly in such alloy systems as discussed in Section 4.2.3., attention must be given to the potential for stress corrosion cracking. Fabrication stresses, such as those associated with welding can be stress relieved by heat treating. In some cases, an entire vessel can be stress relieved after construction (or repair) to reduce the chance of stress corrosion cracking.

A potentially disastrous failure mode is brittle fracture. This condition occurs at low nominal stress and often starts from a defect in the steel, which has become corroded or embrittled. Embrittlement and/or stress corrosion can be caused by chemical factors, such as hydrogen, caustic and chloride ions. It can also be caused by very cold temperatures. These forms of embrittlement are relatively well known, and precautions are often taken in boilers and other process equipment to prevent embrittlement. However, brittle fracture can also occur at ambient or slightly lower than ambient temperatures if a steel component undergoes a ductile-to-brittle transition. This transition is likely to occur more often in heavy-walled vessels such as vessels designed to withstand high pressures, than in thin-walled vessels, which tend to yield rather than fracture. If a vessel contains an undetected defect that could lead to FBL, the failure may occur during the hydrotest when the vessel is stressed to levels greater than the design levels for normal service. For vessels constructed of steels, whose resistance to brittle fracture at low temperature has not been enhanced, it is important that the use conditions are those of the LBF mode. Also, hydrostatic or pneumatic testing conditions should be checked to assure that they also are in the LBF regime.

By the use of fracture mechanics, it is possible to determine whether a vessel is operating in the low-risk LBF range or in the high-risk FBL range. This subject has been addressed in Section 4.2.3, which gives material guidelines for API 620 tanks. Anyone who has critical vessels storing HTHMs is urged to investigate the subjects of fracture mechanics and failure mode diagrams and their applicability to those vessels.[3]

4.3.2 Materials Guidelines (ASME)

For corrosive chemicals, materials of construction must be carefully chosen for compatibility. The considerations discussed in Section 4.2.3 for API 620 tanks also apply to ASME tanks. Where acid gases are likely to be in the air, special external coatings may be necessary. The reader is referred to Appendix E of ASME Vessel Code, "Suggested Good Practice Regarding Corrosion Allowance," for additional information.

Materials of construction for vessels for chemicals that can be mechanically refrigerated or can auto-refrigerate upon sudden release of pressure should be suitable for the low temperature service.

4.4 Additional Guidelines for All Storage Containers

The following are engineering safeguard practices for HTHMs storage designed under API STD 620 or the ASME Pressure Vessel Code.

- In all phases, quality control and materials control should be maintained and documented.
- The minimum wall thickness for any ferrous vessel should be 0.25 inch.
- All nozzles should preferably be at least Class 300 flanged welding necks; when design conditions permit use of Class 150 flanges, these should be used only with blowout resistant gaskets.
- All nozzles of nominal pipe size (NPS) 2 inches and smaller should have a maximum protection of 6 inches. Where insulation requires a greater projection, provision should be made for additional strength or protection. For carbon steel vessels, nozzle wall thicknesses should be at least extra strong (XS).
- Nozzles of 1-1/2 inch and smaller size should be avoided or protected against possible failure caused by falling objects, or other similar loads, by a reinforced steel shield welded to the vessel above the projecting nozzle. Gusseted nozzle reinforcements have been used, but are subject to stress concentrations from flexing loads.
- All nozzle-to-shell welds should be full penetration welds.
- For metals, a limiting weld hardness should be established. For example, for carbon steel, the weld hardness might be limited to a maximum of 225 Brinnell. Hardness should be checked at least every 25 feet for every welder. If hard spots are found, the extent of the welding that is unsatisfactory and the extent of the hard spots should be determined. The hard welds should be ground out, repaired, and rechecked.
- For all welded vessels, hardness tests on welds can be used to assure use of correct welding rod materials.
- All welds should be visually examined.

- Acoustic emission testing is frequently used for testing vessels. A convenient time for its use is during hydrostatic testing.
- Gaskets should be blowout-resistant and fire resistant. Acceptable gaskets include stainless steel, spiral-wound filled with flexible graphite, and ring joint.
- Unshaded, unrefrigerated vessels should be painted white or some other reflective color to minimize the temperature increase on a sunny day. Materials that retain their reflectivity, such as stainless steel, or any other material that has low absorptivity, may remain unpainted except where protection against external corrosion is necessary. If these materials are in locations where ambient conditions can alter absorptivity, then they should also be painted.
- Vessels and vessel supports for flammable, toxic materials should be protected against external fire exposure.

4.5 Piping Design

Piping for HTHMs, as covered in this document, falls within the scope of Chapter VIII of the Pressure Piping Code, "Piping for Category M Fluid Service." Category M fluid service is defined as "a fluid service in which the potential for personnel exposure is judged to be significant and in which a single exposure to a very small quantity of a toxic fluid, caused by leakage, can produce serious irreversible harm to persons on breathing or bodily contact, even when prompt restorative measures are taken." This is similar but not exactly the same definition as that for "Lethal Service" of the ASME Pressure Vessel Code. As with the Pressure Vessel Code, the owner has overall responsibility for compliance with the Pressure Piping Code and for establishing the requirements for design, including the determination of whether the fluid service is Category M.

4.6 Additional Guidelines for Piping Systems

Piping systems should be "safeguarded" as required by the Code and as outlined in Appendix G. Safeguarding by plant layout is covered in other chapters. The Guidelines in this section are engineering safeguard practices that should be applied in design, construction, and fabrication of piping systems for HTHMs.

4.6.1 Design Guidelines

Piping systems should meet the requirements for both Category M fluid service and for "severe cyclic conditions." As required by the Code, piping systems should be subjected to a flexibility analysis and, if found to be too rigid, flexibility should be added. In addition, damping or stabilization of fluid flow dynamics should be used to eliminate or minimize destructive loads, such as severe vibration pulsations or severe cyclic conditions. Expansion bellows, flexible connection, and glass viewports or "bulls eyes" may be weakened or broken by vibration, stress, and cyclic flow. These devices should be avoided and, where unavoidable, should be subjected to the most stringent administrative control, that is, listed and inspected even more frequently than regularly required. Pipelines should be designed with the minimum number of joints, fittings, and valves. Joints in piping systems for highly toxic chemicals should be flanged or butt-welded. Threaded joints should not be used because threading reduces the thickness of the pipe and introduces a notch at which fatigue or fracture might begin. Threaded joints are also prone to leakage and are particularly bad in corrosive service. Threaded joints are permitted for off-line instruments, such as pressure gauges, where a failed joint can be isolated quickly with a block valve.

A flow-limiting orifice may be inserted in the pipe to limit the rate of release. This is effective when large flows are not required through the line under normal conditions, for example, restriction orifices are often used in compressed gas cylinder or instrument piping. This allows the use of larger, heavier-walled pipe in a vulnerable situation without increasing the flow of material.

Piping systems should be pressure-protected if sections can be isolated with an expandable liquid trapped between valves. Where this is possible, overpressure protection must be installed on the pipe-line between the valves. This protection may take the form of a pressure relief valve, an expansion chamber, or a rupture disk to a closed pressure-alarmed tank. Since these protective devices also create possible sources of leakage or discharge, the number of shutoff valves installed should be optimized for safe and efficient operation.

When the freezing point of the material in a pipe is more than a few degrees above the lowest expected ambient temperature, precautions (e.g., heat tracing) are necessary to guard against freezing. Even if the material does not expand on freezing, such precautions are necessary, because then the material may thaw unevenly, which could lead to excessive hydrostatic pressure from liquid expansion.

Thermal insulation, shields, process controls, internal cooling and external spray protection are all ways to protect against excessively high temperatures resulting from internal or external conditions. Excessively high temperatures usually result from fire or exothermic reactions. Precautions against exothermic reactions require control of quality and conditions of material in storage; control of external heating, especially that from fire exposure; and control of internal heating by proper process controls and internal cooling. Fire protection, spray protection and diking are covered in the AIChE-CCPS "Guidelines for Vapor Release Mitigation." When there is a possibility of exothermic reactions, temperature monitoring of the system is necessary (see Chapter 6, which covers instrumentation).

Where there is fire potential, piping and components should have a fire-resistance capability of at least 30 minutes at 1100° F. The melting point, degradation temperature, loss of strength at elevated temperature, the susceptibility to brittle failure, or failure from thermal shock of the materials when exposed to fire or to fire-fighting measures, and even the possible combustibility of the piping material and its reactivity with the lading under high temperature exposure are all important.

If the properties of the HTHM being handled are such that there are no satisfactory materials of construction with adequate fire resistance, insulation should be used to keep the piping from overheating during a fire. In selecting an insulation for that purpose, the insulation stability, fire resistance, and ability to remain in place during a fire should be considered. The problem of hidden corrosion under the insulation in corrosive atmospheres should also be recognized. Spray protection may be an alternative to insulation. In other cases, both spray and insulation might be used.

It is preferred engineering practice to have fill and discharge lines enter tanks through the top. While, for liquids near their boiling point, this requires the use of submerged pumps for tank discharge, it avoids a piping penetration below the liquid level in the tank. However, servicing of submerged pumps in HTHM service also presents some problems. Siphoning effects have to be considered in designing such fill and discharge systems.

Geysering is a phenomenon that can result in a destructive pressure surge and should be guarded against. Geysering may occur in piping that handles fluids at or near their boiling temperatures under conditions when the rapid evolution of vapor within the piping causes a rapid expulsion of liquid. Geysering is usually associated with vertical pipelines, but it may sometimes occur in inclined lines as well.

Another effect that may cause unacceptable stresses at anchors, supports, guides, and within pipe walls is bowing during cooldown. This can occur, usually in horizontal piping, on the introduction of a refrigerated fluid, at a flow rate that allows stratified two-phase flow, causing large temperature gradients.

Two-phase flow can also generate excessive pressure oscillations and surges that may damage the piping. Analysis for water hammer caused by rapid valve closure should be conducted.

When absolutely no leakage can be tolerated in a piping system, consideration should be given to a piping system designed with (a) pipes within pipes with the annular space filled with an inert gas at a higher pressure than the toxic gas so that leakage is inward or with the inert gas at a lower pressure with continuous analysis of the inert gas for leakage out; or (b) single pipes with flanges with double seals and valves with bellow seals and a continuous flow of inert purge gas between the flange seals and through the bellows to a continuous analyzer that monitors for leakage. Variations of such systems have been successfully used for pyrophoric gases that cannot be allowed to leak out, and for air-reactive materials where no inward leakage of air can be allowed.

4.6.2 Construction Guidelines

Special attention to cleanliness in fabrication, assembly, erection, testing, inspection, and post-erection cleaning is essential when contamination must be avoided, as when accommodating strong oxidizers and other highly reactive chemicals, for example, oxygen, chlorine, or fluorine. Attention to cleanliness is also necessary in low-temperature service, where moisture, grease, oil, and other contaminants have to be removed to prevent sticking of valves and blockage of components.

To reduce the effects of differential thermal expansion and the likelihood of stress relaxation and loss of bolt tension in flanged joints, the use of controlled bolting procedures are recommended for Category M fluid service, especially in high, low, or cycling temperature services, or under conditions involving vibration or fatigue.

Access is required for periodic inspections, and good visibility and color coding provide easy identification. In some cases, leak-detecting paint and/or guards may be applied to flanges.

If the piping system requires special cleaning because of the properties of the fluid being carried, pockets should be avoided and adequate vent and drain connections should be provided to introduce and remove cleaning, rinsing, and drying materials. Wherever possible, blind flanges rather than valves should be used on the vents and drains.

When a highly toxic chemical is also highly corrosive, or the environment in the area is corrosive, care should be taken in the selection or protection of the supports. Field-specified supports should meet all support design requirements. Hanger types supported from rods should generally not be used because of the possibility of aggressive corrosion quickly attacking the support rods. Hangers supporting the pipe from underneath are recommended. If necessary, pipe supports should be protected from extremes of temperature (potential process and climatic excursions or external fire), since failure may result from exposure to low temperatures (which may embrittle the supports) or high temperatures (which may weaken them).

4.6.3 Materials Guidelines

For corrosive chemicals, materials of construction must be carefully chosen for compatibility. The corrosion allowance provided in the design should be adequate for the projected life of the piping. External corrosion can also endanger the piping. Piping that may be subjected to salt spray from the ocean, or operated at temperatures where condensation is likely, is particularly vulnerable. Special attention should be paid to the areas around supports. Where acid gases are likely to be in the air, special external coatings may be necessary.

Materials of construction must be resistant to the physical abuse that can reasonably be expected. A properly supported NPS 1-1/2 inch steel pipe is generally accepted as being resistant to loads, such as when being stepped on or struck by a heavy pipe wrench. Pipelines of other materials should be at least as tough as 1-1/2 inch pipe with extra strong wall thickness, and they should be protected from physical damage.

As required by the Code, brittle materials such as cast iron and plastics should not be used. Ductile cast and malleable iron are also avoided by most companies for Category M service because of the possibility of inadvertently receiving cast iron components instead of ductile ones.

Where the piping will operate at low temperatures, or be subject to auto-refrigeration and the chilling effect of a sudden loss of pressure on a highly volatile fluid, the piping system and its components should be made of impact-resistant materials suitable for the minimum temperature to which they could be exposed.

Gasketing should be resistant to blowout to prevent a small leak from escalating to a larger release of material through a flange. Metal reinforced, spiral-wound, and ring joint gaskets of a material suitable for the temperature and the fluid being handled are considered blowout-resistant.

Gaskets, packing, seals, lubricants, and insulation should be carefully chosen for compatibility with the fluid service, such as strong oxidizers, concentrated solvent, strong acid, strong caustic, or otherwise highly corrosive chemicals.

4.6.4 Routing Guidelines

Pipe lines should be routed in protected areas away from traffic lanes. Armoring of the piping, providing guards for nozzles and small size piping, and/or providing barricades should be considered for protecting the piping system against physical abuse. Consideration should also be given to the possible consequences of exposure to fire when routing the lines near storage, pumps, or pipelines that handle flammable fluids.

4.6.5 Valve Guidelines

Remotely operated block valves located adjacent to the tanks should be provided to limit the quantity of highly toxic material that could escape as a result of pipe ruptures or leaks in fittings, packing, or seals.

Remotely controlled, quick-closing (be aware of water hammer effects), close-coupled (bolted directly to the tank nozzle flange) valves should be provided on all lines that can discharge liquid from the tank, and considered for larger vapor lines on pressurized storage tanks. These emergency block valves should be designed to be fail-safe, typically normally closed, and be open only when the tank is operating. When the toxic chemical is also flammable, these valves should close automatically in the event of fire and be fire-resistant. Plug valves may not be satisfactory for automated block valves, because an unpredictably high torque may be required when valves are adjusted to stop leakage. Flanged, top-entry ball valves and high-pressure butterfly valves are recommended.

Powered operators or actuators should be provided for valves whose function is to limit flow during an emergency. Actuators should be mounted in such a way that a mechanic will not mistakenly loosen the pressure-containment bolting in order to remove the actuator. Power actuated valves should be capable of being operated manually. Open-ended external valves should be provided with blind flanges or plugs.

Because of the expansion characteristics of some liquefied gases, such as chlorine, and materials that can polymerize, only valves that cannot trap liquid when closed should be used. Acceptable valves include globe, ball with a pressure relief to the high-pressure side, plug with pressure relief to the high-pressure side, and flanged, high-pressure butterfly. Design must be based on the assumption that no valve will isolate perfectly in service, and the layout must provide for the convenient insertion of slip blinds for effective isolation. If slip blinds are not used, double block-and-bleed systems should be provided at critical points. This system consists of two valves in series with provision for venting or evacuating the space between.

Where difficulty may be encountered in maintaining valves in an effective condition, piping design should anticipate the need for periodic insertion of slip blinds so that valves can be removed for cleaning and maintenance.

The type of valve chosen for a particular service should be such that the packing gland on the stem seals effectively. Packing shrinkage is possible in low-temperature fluid service. To avoid packing shrinkage and external icing or other heat flux problems, extended bonnet valves are recommended, where necessary, to establish a temperature differential between the valve stem packing and the fluid in the piping.

When leakage around the valve stem cannot be tolerated, valves with double packing boxes or with a bellows seal (rather than conventional packing) can be used.

Polymerizable materials present some unusual limitations in the selection of valves. When rodding is necessary to unplug a line, a gate or ball valve may be preferable to a globe valve, because the latter cannot be rodded through and will also restrict free drainage in a horizontal line. Gate valves, on the other hand, have a stagnant zone in the bonnet in which liquid can be trapped, and there have been cases of gate valves blowing apart. Engineering judgment is necessary in making a choice. Finally, bellows-sealed valves have a cavity into which a material can seep and polymerize, unless the bellows is continuously purged.

4.6.6 Pump Guidelines

The choice of pumps for a particular HTHM storage containment system is dependent on the properties of the material and the layout of the system. The design of the pumping system falls outside the scope of this document. However, the following aspects merit emphasis.

Suction lines for pumps should generally be kept as short as possible to avoid high pressure drop in the line and to assure an adequate net positive suction head for the pump. The inlet line should be provided with an emergency block valve designed to shut off remotely and/or automatically when there is an uncontrolled release of a toxic material or a fire.

The pump should be constructed of materials that are compatible with the HTHM being handled. The design pressure rating of the pump should not be less than the pressure at the most severe condition of coincident (internal or external) pressure and (minimum or maximum) temperature expected during service. For refrigerated liquefied gases or corrosive materials, the choice of a pump constructed of the proper materials is of primary concern because of the possibility of low-temperature embrittlement, stress corrosion cracking, or brittle fracture.

The pressure parts of the pump should not be constructed of cast, wrought, malleable, or ductile iron because of their lack of ductility or their sensitivity to thermal or mechanical shock. Since the stuffing box on an ordinary centrifugal pump is a primary source of leakage -- and even a mechanical seal will leak -- special pumps should be used with highly toxic materials. Seal-less or gland-less pumps, often called canned pumps, should be considered. In a canned pump, the impeller of the pump and the rotor of the motor are mounted on an integral shaft, the rotor and stator being encased or canned, with the chemical circulating in the space between the stator and the impeller/rotor. To prevent fugitive emissions, a double mechanical seal pump, when properly maintained, runs with virtually no fugitive leakage. The double mechanical seal can be designed to run at higher or lower pressure than the process piping, based upon the specific materials and contamination sensitivities. In addition, the space between the seals can be connected to an instrumented seal pot with both pressure and level alarms to detect failure of either seal before a catastrophic failure occurs. Alternatively, a single mechanical seal with packing gland backup also provides protection against catastrophic failure.

Serious releases of toxic materials can result if a pump is destroyed by improper operation. Two of the most serious hazards are deadheading and cavitation. Deadheading is a condition that occurs when a liquid-filled pump is running with a blocked outlet piping, or even worse, with both inlet and outlet blocked. Cavitation occurs when a fluid begins incipient vaporization within the pump and the resultant bubbles collapse on the pump impeller as pressure-induced condensation occurs. In the case of deadheading, the pump can generate enough heat and/or pressure to destroy itself if the condition persists. In several cases involving reactive (and toxic) materials, internal runaway decompositions or reactions between the pump metal and the material (e.g., iron/chlorine) have occurred within the pumps, destroying them catastrophically.

Cavitation can be prevented by providing an adequate liquid head on the suction side of the pump. Deadheading can be prevented by proper operation, but should be supplemented by engineering controls. Instrumentation can be used to detect rises in temperature and/or pressure and to shut off the pump. Finally, a recirculating bypass without valves from the pump discharge back to the pump suction or the source of the pump suction, such as a tank, can be installed. This type of safeguard against deadheading should be used with caution if a highly toxic material that is also reactive can be contaminated by materials in the discharge piping or overheated by the recirculation. Overheating is more likely to occur if the recirculation goes directly to the pump suction.

References

1. Lees, F. P., "Loss Prevention in the Process Industries," Vol. I, p. 277ff; Vol. II, p. 749ff, Butterworth, Boston and London, 1980.
2. Perry, R. H. and D. Green, "Chemical Engineers' Handbook," 6th ed., p. 23-4, McGraw-Hill, New York, 1984.
3. Rolfe, S. T. and J. M. Barson, "Fracture and Fatigue Control in Structures: Applications of Fracture Mechanics," Prentice-Hall, New Jersey, 1977.

TABLE 4.1
ASME CODES

American Society of Mechanical Engineers
345 East 47th Street
New York, NY 10017.

ASME Boiler and Pressure Vessel Code, 1986

Section	I	Power Boilers
	II	Materials Specification
	III	Division 1 - Nuclear Power Plant Components
		Division 2 - Code for Concrete Reactor Vessels and Containments
	IV	Heating Boilers
	V	Non-Destructive Testing
	VI	Recommended Rules for Care and Operation of Heating Boilers
	VII	Recommended Rules for Care of Power Boilers
	VIII	Pressure Vessels Division 1
		Pressure Vessels Division 2 - Alternative Rules
	IX	Welding and Brazing Qualifications
	X	Fiberglass - Reinforced Plastic Pressure Vessels
	XI	Rules for In-service Inspection of Nuclear Power Plant Components Code Cases Book

ASME Code for Pressure Piping (B31)

B31.1	1986	Power Piping
B31.2	1968	Fuel Gas Piping
B31.3	1987	Petroleum Refinery Piping
B31.4	1986	Liquid Petroleum Transportation Piping Systems
B31.5	1987	Refrigeration Piping
B31.8	1986	Gas Transmission and Distribution Piping Systems

Performance Test Codes

PTC 6.1 - 1984	Steam Turbines
PTC 7.1 - 1969	Reciprocating Steam-Driven Displacement Pumps
PTC 8.2 - 1973	Centrifugal Pumps
PTC 10 - 1986	Compressors and Exhausters, 1973
PTC 12.2 - 1983	Steam Condensing Apparatus
PTC 19.2 - 1964	Pressure Measurement
PTC 19.3 - 1985	Temperature Measurement
PTC 19.13 - 1961	Measurement of Rotary Speed
PTC 22 - 1985	Gas Turbine Power Plants, 1973
PTC 25.3 - 1976	Safety and Relief Valves with Atmospheric or Superimposed Back Pressure Before Discharging
PTC 31 - 1985	Ion Exchange Equipment

General and Safety Standards

A13.1 - 1985	Scheme for the Identification of Piping Systems
A112.26.1 - 1984	Water Hammer Arresters, 1975
B15.1 - 1984	Safety Standard for Mechanical Power Transmission Apparatus, 1972
B19.3 - 1986	Safety Standard for Compressors for Process Industries
B20.1 - 1984	Safety Standards for Conveyors and Related Equipment
B30.9 - 1984	Slings
B30.16 - 1981	Overhead Hoists
B36.10 - 1985	Welded and Seamless Wrought Steel Pipe
B36.19 - 1985	Stainless Steel Pipe, 1971
B56.1 - 1983	Low Lift and High Lift Trucks
B73.1 - 1984	Specifications for Horizontal, End Section Centrifugal Pumps for Chemical Process

---

TABLE 4.2
API PUBLICATIONS

American Petroleum Institute
1220 L Street, N.W.
Washington, DC 02005.

Guide for Inspection of Refinery Equipment

Ch. 1	Introduction, 2nd ed., 1976
2	Conditions causing deterioration or failures, 2nd ed., 1973
3	General preliminary and preparatory work, 2nd ed., 1976
4	Inspection tools, ed., 1983
5	Preparation of equipment for safe entry and work, ed., 1978
6	Unfired pressure vessels, ed., 1982
7	Heat exchangers, condensers and cooler boxes, 2nd ed., 1967
8	Direct-fired boilers and auxiliary equipment, 2nd ed., 1974
9	Fired heaters and stacks, ed., 1981
10	Pumps, compressors and blowers, and their drivers, 2nd ed., 1976
11	Pipes, valves and fittings, 2nd ed., 1974
12	Foundations, structures and buildings, 2nd ed., 1975
13	Atmospheric and low-pressure storage tanks, 4th ed., 1975
14	Electrical systems, 3rd ed., 1982
15	Instruments and control equipment, 3rd ed., 1981
16	Pressure-relieving devices, 3rd ed., 1985
17	Auxiliary and miscellaneous equipment, 2nd ed., 1978
18	Protection of idle equipment, 3rd ed., 1982
	Appendix inspection of welding, ed., 1985

Manual on Disposal of Refinery Wastes
Volume on liquid wastes, in preparation
Volume on solid wastes, 1st ed., 1977
Volume on atmospheric emissions, 1st ed., 1977

Standards

Std 510	Pressure Vessel Inspection Code - Maintenance Inspection, Rating, Repair Alteration, 4th ed., 1985
Std 526	Flanged steel safety relief valves, 3rd ed., 1984
527	Commercial seat tightness of safety relief valves with metal-to-metal seats, 2nd ed., 1982
598	Valve inspection and test, 5th ed., 1982
600	Steel gate valves, flanged or buttwelding ends, 8th ed., 1981
605	Large-diameter carbon steel flanges, 3rd ed., 1980
610	Centrifugal pumps for general refinery services, 6th ed., 1981
611	General-purpose steam turbines for refinery services, 2nd ed., 1982
612	Special-purpose steam turbines for refinery services, 2nd ed., 1979
613	High-speed, special-purpose gear units for refinery services, 2nd ed., 1977
614	Lubrication, shaft-sealing and control oil systems for special-purpose applications, 2nd ed., 1984
615	Sound control of mechanical equipment for refinery services, 1973
616	Combustion gas turbines for general refinery services, 2nd ed., 1982
617	Centrifugal compressors for general refinery services, 4th ed., 1979
618	Reciprocating compressors for general refinery services, 2nd ed., 1974
619	Rotary-type positive displacement compressors for general refinery services, 2nd ed., 1985

(continued on p. 38)

TABLE 4.2 (Cont.)

620	Recommended rules for design and construction of large, welded, low pressure storage tanks, 7th ed., 1982 and Revision 1, 1985
650	Welded steel tanks for oil storage, 7th ed., 1980 and Revision 1, 1984
660	Heat exchangers for general refinery services, 4th ed., 1982
661	Air-cooled heat exchangers for general refinery services, 2nd ed., 1978
1104	Standard for welding pipelines and related facilities, 16th ed., 1983
2000	Venting atmospheric and low pressure storage tanks, 3rd ed., 1982
2510	Design and Construction of LP-gas installations at marine and pipeline terminals, natural gas processing plants, refineries and tank farms, 5th ed., 1985
Recommended Practices	
RP 50	Recommended gas plant good operating practices for protection of the environment, 1975
500A	Recommended practice for classification of areas for electrical installations in petroleum refineries, 4th ed., 1982
500C	Recommended practice for classification of areas for electrical installation at petroleum and gas pipeline transportation facilities, 2nd ed., 1984
520	Recommended practice for the design and installation of pressure relieving systems in refineries Pt 1 Design, 4th ed., 1976 2 Installation, 2nd ed., 1963
521	Guide for pressure relief and depressuring systems, 2nd ed., 1982
530	Recommended practice for calculation of heater tube thickness in petroleum refineries, 2nd ed., 1978
540	Recommended practice for electrical installations in petroleum processing plants, 2nd ed., 1974
541	Recommended practice for form-wound squirrel-cage induction motors, 200 h.p. and larger, 1972
550	Manual on installation of refinery instruments and control systems Pt 1 Process instrumentation and control, Sections 1-14, 1977-1985 2 Process stream analyzers, Section 1-11, 1983-1985 3 Fired heaters and inert gas generators, 3rd ed., 1985 4 Steam generators, 2nd ed., 1984
1102	Recommended practice for liquid petroleum pipelines crossing railroads and highways, 5th ed., 1981
1107	Recommended pipeline maintenance welding practices, 2nd ed., 1978
1109	Recommended practice for marking liquid petroleum pipeline facilities, 2nd ed., 1986
1110	Recommended practice for the pressure testing of liquid petroleum pipelines, 2nd ed., 1981
2001	Fire protection in refineries, 6th ed., 1984
2003	Protection against ignitions arising out of static, lightning and stray currents, 4th ed., 1982
2009	Safe practices in gas and electric cutting and welding, 3rd ed., 1967
2013	Cleaning mobil tanks in flammable or combustible liquid service, 5th ed., 1975
2015	Cleaning petroleum storage tanks, 3rd ed., 1985

TABLE 4.3
SELECTED ANSI STANDARDS

American National Standards Institute
1430 Broadway
New York, NY 10018.

B	Mechanical
B1	Screw Threads
B16	Pipe Flanges and Fittings
B18	Bolts and Nuts
B31	Pressure Piping
B31.1-1973	Power Piping
B31.2-1968	Fuel Gas Piping
B31.3-1973	Petroleum Refinery Piping (with Addenda)
B31.4-1974	Liquid Petroleum Transportation Piping Systems
B31.5-1974	Refrigeration Piping
B31.8-1975	Gas Transmission and Distribution Piping Systems
ANSI/CGA-V-1-1977	Compressed Gas Cylinder Valve Outlet and Inlet Connections
B73.IM-1984	Specifications for Horizontal End Suction Centrifugal Pumps for Chemical Processes
B78.1-1982	Tubular Heat Exchangers in Chemical Process Service
B95.1-1977	Terminology for Pressure Relief Devices
B96.1-1986	Specification for Welded Aluminum-Alloy Field-Erected Storage Tanks
ANSI-UL-668-1982	Safety Standard for Hose Valves for Fire Protection Service (UL 668-May 1970)
ANSI/UL-448-1985	Safety Standard for Pumps for Fire Protection Service (UL 448-February 1975)
ANSI/UL-569-1982	Safety Standard for Pigtails and Flexible Hose Connections for LP-Gas
ANSI/UL-58-1985	Steel Underground Tanks for Flammable and Combustible Liquids
ANSI/API-2510-1985	Design and Construction of LP-Gas Installations at Marine and Pipeline Terminals, Natural Gas Processing Plants, Refineries and Tank Farms
ANSI/UL-536-1984	Flexible Metallic Hose
ANSI/UL-644-1981	Safety Standard for Container Assemblies for LP-Gas
ANSI/UL-526-1984	Flanged Steel Safety Relief Valves
ANSI/UL-567-1983	Safety Standard for Pipe Connectors for Flammable Liquids and LP-Gas
ANSI/UL-331-1984	Safety Standard for Strainers for Flammable Fluids and Anhydrous Ammonia
ANSI/UL-80-1980	Safety Standard for Liquid-Level Gauges and Indicators for Petroleum Products
ANSI/UL-51-1980	Power-Operated Pumps for Anhydrous Ammonia and LP-Gas
ANSI/API-620-1986	Rules for Design and Construction of Large, Welded, Low-Pressure Storage Tanks
ANSI/API-650-1980	Welded Steel Tanks for Oil Storage
ANSI/UL-467-1984	Safety Standard for Grounding and Bonding Equipment
ANSI/UL-886-1985	Outlet Boxes and Fittings for Use in Hazardous Locations
ANSI/UL-844-1984	Safety Standard for Electric Light Fittings for Use in Hazardous Locations
ANSI/UL-894-1985	Safety Standard for Switches for Use in Hazardous Locations
ANSI/UL-698-1985	Safety Standard for Industrial Control Equipment for Use in Hazardous Locations
ANSI/UL-823-1985	Safety Standard for Electric Heaters for Use in Hazardous Locations
ANSI/UL-45-1986	Safety Standard for Portable Electric Tools

(continued on p. 40)

TABLE 4.3 (Cont.)

ANSI/UL-674-1984	Safety Standard for Electrical Motors and Generators for Use in Hazardous Locations, Class I, Groups C and D
ANSI/UL-508-1983	Safety Standard for Industrial Control Equipment
ANSI/UL-674-1984	Safety Standard for Electric Motors and Generators for Use in Hazardous Locations, Class II, Groups E, F and G
ANSI/NEMA-MG2-1983	Safety Standard for Construction and Guide for Selection, Installation and Use of Electric Motors and Generators
K13.1-1973	Identification of Air-Purifying Respirators Canisters and Cartridges
K61.1-1981	Safety Requirements for Storage and Handling of Anhydrous Ammonia
ANSI/ISA-S26-1968	Dynamic Response Testing of Process Control Instrumentation
ANSI/ISA-RP-55.1-1975	Hardware Testing of Digital Process Computers
MH5.1.3M-1982	Requirements for Tank Containers for Liquids and Gases
Z9.1-1977	Practices for Ventilation and Operation of Open-Surface Tanks
Z9.2-1979	Fundamentals Governing the Design and Operation of Local Exhaust Systems
ANSI/UL-525-1984	Flame Arresters for Use on Vents of Storage Tanks for Petroleum Oil and Gasoline
ANSI/NFPA-495-1985	Explosive Materials; Manufacture, Transportation, Storage and Use of

TABLE 4.4
SELECTED NFPA STANDARDS

National Fire Protection Association
Batterymarch Park
Quincy, MA 02269.

NFPA11	1983	Foam Extinguishing Systems
11A	1983	High Expansion Foam Systems
12	1985	Carbon Dioxide Extinguishing Systems
12A	1985	Halogenated Agent Extinguishing Systems - Halon 1301
12B	1985	Halogenated Fire Extinguishing Agent Systems - Halon 1211
13	1987	Installation of Sprinkler Systems
13A	1987	Maintenance and Care of Sprinkler Systems
14	1986	Standpipe and Hose Systems
15	1985	Water Spray Fixed Systems
16	1986	Foam-Water Sprinkler and Spray Systems
16A	1983	Closed-Head Foam-Water Sprinkler Systems
17	1985	Dry Chemical Extinguishing Systems
20	1987	Centrifugal Fire Pump
30	1984	Flammable and Combustible Liquids Code
43A	1980	Storage of Liquids and Solid Oxidizing Materials
43C	1986	Code for Storage of Gaseous Oxidizing Materials
49	1975	Hazardous Chemicals Data
50	1985	Bulk Oxygen Systems at Consumer Sites
50A	1984	Gaseous Hydrogen Systems at Consumer Sites
50B	1985	Liquefied Hydrogen Systems at Consumer Sites
51B	1984	Fire Prevention in Use of Cutting and Welding Processes
53M	1985	Fire Hazards in Oxygen-Enriched Atmospheres
54	1984	National Fuel Gas Code
58	1986	Liquefied Petroleum Gases, Storage and Handling
59A	1985	Production, Storage and Handling of Liquefied Natural Gas
68	1978	Explosion Venting Guide
69	1986	Explosion Prevention Systems
70	1987	National Electrical Code
70B	1983	Electrical Equipment Maintenance
72E	1984	Automatic Fire Detectors
75	1987	Electronic Computer/Data Processing Equipment
77	1983	Recommended Practice on Static Electricity
78	1986	Lightning Protection Code
101	1985	Life Safety Code
1962	1979	Care, Maintenance and Use of Fire Hose
214	1983	Water Cooling Towers
231C	1986	Rack Storage of Materials
232	1986	Protection of Records
306	1984	Control of Gas Hazards on Vessels to be Repaired
307	1985	Marine Terminals, Piers and Wharves
321	1987	Basic Classification of Flammable and Combustible Liquids
325M	1984	Fire Hazard Properties of Flammable Liquids, Gases and Volatile Solids
329	1987	Underground Leakage of Flammable and Combustible Liquids
385	1985	Tank Vehicles for Flammable and Combustible Liquids
490	1986	Storage of Ammonium Nitrate

(continued on p. 42)

TABLE 4.4 (Cont.)

491M	1986	Manual of Hazardous Chemical Reactions
493	1978	Intrinsically Safe Apparatus Locations
495	1985	Code for Explosive Materials
496	1986	Purged and Pressurized Enclosures for Electrical Equipment
497A	1986	Classification of Class I Hazardous Locations for Electrical Installations in Chemical Plants
498	1986	Explosives Motor Vehicle Terminals
512	1984	Truck Fire Protection
601	1986	Guard Service in Fire Loss Prevention
602	1986	Guard Operations in Fire Loss Prevention
655	1982	Prevention of Sulfur Fires and Explosions
704	1985	Identification of the Fire Hazards of Materials
901	1986	Uniform Coding for Fire Protection
902M	1986	Fire Reporting Field Incident Manual

TABLE 4.5
OTHER CODES

Chemical Industry Safety and Health Council
Alembic House
93 Albert Embankment
London SE1 7TU

Codes of Practice for Chemicals with Major Hazards: Chlorine
Codes of Practice for Chemicals with Major Hazards: Ethylene dichloride
Codes of Practice for Chemicals with Major Hazards: Ethylene oxide
Codes of Practice for Chemicals with Major Hazards: Hydrogen Chloride (Anhydrous)
Codes of Practice for Chemicals with Major Hazards: Phosgene

Chemical Industries Association, Ltd.
Alembic House
93 Albert Embankment
London SE1 7TU

Code of Practice for the Storage of Anhydrous Ammonia Under Pressure in the United Kingdom, Spherical and Cylindrical Vessels (1980)
Code of Practice for the Large Scale Storage of Fully Refrigerated Anhydrous Ammonia in the United Kingdom (1975)

Chlorine Institute, Inc.
2001 L Street NW
Washington, DC 20036

Non-Refrigerated Liquid Chlorine Storage, 4th ed., (1982)
Refrigerated Liquid Chlorine Storage, 1st ed., (1984)

Compressed Gas Association
1235 Jefferson Davis Highway
Arlington, VA 22202

G-2	Anhydrous Ammonia
G-2.1	Storage and Handling of Anhydrous Ammonia
G-3	Sulfur Dioxide
G-8.1	Nitrous Oxide Systems
G-12	Hydrogen Sulfide
P-13	Safe Handling of Liquid Carbon Monoxide

National Association of Corrosion Engineers
1440 South Creek Drive
Houston, TX 77084

RP-01.70	Protection of Austenitic Stainless Steel from Polythionic Acid Stress Corrosion Cracking During Shutdown of Refinery Equipment
RP-01.72	Surface Preparation of Steel and Other Hard Materials by Water Blasting Prior to Coating or Recoating
RP-01.75	Control of Internal Corrosion in Steel Pipelines and Piping Systems
RP-01.78	Design, Fabrication, and Surface Finish of Metal Tanks and Vessels to be Lined for Chemical Immersion Service
RP-01.81	Liquid Applied Internal Protective Linings and Coatings for Oil Field Production Equipment
RP-01.84	Repair of Lining Systems

5

LOADING AND UNLOADING FACILITIES

Loading and unloading facilities that are often incorporated in handling and storage systems are vulnerable to incidents that can and have resulted in loss of containment. Also, since the current practice at some manufacturing facilities is to transfer materials directly from shipping containers to the process without any intermediate storage, this procedure, while reducing inventory, places increased importance on the design of the unloading facility. In this chapter, guidelines are provided for rail car and tank truck facilities, barge unloading facilities, and areas used for nonpermanent storage.

Many of the general precautions that are discussed earlier apply to loading/unloading facilities, but these facilities present additional problems that are not normally encountered in processing or storage areas. These problems include: fragile connections between transport vessels and fixed facilities, the possibility of moving transport vessels during the transfer operation, and the difficulty of providing secondary containment for releases from transport vessels and transfer facilities.

5.1 Rail Car and Tank Truck Facilities

Since transfer facilities, in general, are relatively high-hazard areas, they should be separated from process areas.[1] In the case of HTHMs, the location of loading/unloading stations should be carefully evaluated using the same Guidelines as for permanent storage areas (see Chapter 3). Proper planning for traffic is also important to the safe operation of terminals by eliminating congestion that may cause spill incidents and sources of ignition that can cause fires. For truck terminals, it is desirable that vehicles be able to drive in and out of the facility without reversing.[1] For rail cars, loading/unloading should be performed on dedicated sidings.

The layout design should anticipate the handling of spills. The pavement at the spotting area should slope away from the loading rack, particularly when flammable liquids are to be handled, so that the spill does not collect under the vehicle. The area should also be curbed to prevent the spread of spills to other areas. Use of a collection sump or foam-dispensing system can be effective in reducing vapor generation by minimizing the surface area exposed to atmosphere. There should be no open drains in the curbed area by which material could be transferred to other plant areas in an uncontrolled or unpredictable manner.

The transfer of material to and from shipping containers is a potentially hazardous activity; however, risks of serious incidents can be reduced by the incorporation of proper design features. The transfer operation should be under the control of the owner's attendant.

Before making connections to tank trucks or rail cars, provisions should be taken against accidental movement by means of wheel chocks on trucks and rail cars. For rail cars, derailers should be installed between the car and other rail lines to protect the rail car from other cars being moved.

Quality control procedures should be established to confirm the identity of materials before unloading. Unloading of incompatible materials can result in undesirable reactions in storage tanks or process vessels. Also, before loading, the heel in the container to be loaded should be identified. Use of dedicated shipping containers with standardized connections is definitely preferred. Common practice is to use different types of connectors, or different diameter lines, to avoid loading or unloading incompatible materials into the wrong tank. This approach, however, should be used in addition to, but not in place of, a vigorous quality control program.

After identifying the material to be transferred, the operator should confirm that the receiving tank or vessel has sufficient free volume to accept the entire contents of the shipping container or the appropriate transfer quantity. All safeguards for monitoring transfer operations should be in place and operable.

Current good practice would design for the transfer of material from top connections of rail cars and tank trucks. For high vapor pressure HTHM, a dual vapor/liquid transfer line system should be employed, with one line dedicated to transfer of liquid to and from the container and the other line used for venting the container during filling. The vent line can return to the container being unloaded or communicate with a vent scrubber designed to neutralize toxic vapors or with a vapor recovery system.

For a HTHM, a padding gas is sometimes used as a motivating force for unloading shipping containers. Some typical padding gases include dry air, nitrogen, or the vapor from the material being transferred. If the supply pressure for displacement gas is above the set pressure of the relief valve on the shipping container, a relief valve should be provided on the gas line set below the relief valve set pressure of the shipping container. Protection against backflow of a HTHM into the padding gas should also be provided. When nitrogen is used, it should be of a quality that assures compatibility with the HTHM. Nitrogen from an air separation unit, or from a liquid nitrogen storage/vaporization supply system usually ensures dry gas with low oxygen content. Air should be oil-free and dry (e.g., minus 40° F dew point for chlorine unloading).[2] When considering the use of air, compatibility with the particular HTHM is extremely important.

The use of a vaporized HTHM liquid as a pad gas requires special care to avoid overheating as well as increased potential for corrosion. The choice of a heating medium is very important in limiting local surface temperature in the vaporizer. For example, chlorine vaporizers almost always use hot water as the heat-transfer medium to avoid overheating. Good controls are essential for systems of this type. Electrically heated vaporizers are generally to be avoided. Caution is also advised in the use of a vaporized liquid when the material is thermally sensitive or prone to polymerization.

A compatible gas system for purging unloading lines should be available to avoid small releases while disconnecting lines. The purge gas should be treated (e.g., in a vent scrubber) before release to the environment. A catch tank can be provided upstream of the treater to trap and vaporize expelled liquid; or, where practical, the trapped liquid may be recovered.

Connection systems involving flexible hoses, articulated arms, and hard piping are all commonly used. Careful consideration should be given to the specification of design pressure and materials of construction for the connector selected. Articulated arms have advantages over hoses in terms of better reliability, being more durable and easier to hook up. Arms have a tendency to develop small leaks at the joints, while hose failures usually result in larger releases. For this reason, if hoses are used, it is critical that they be dedicated to and compatible with the HTHM, as well as being properly designed, identified, tested, and protected from all sources of damage. In either case, dry-break systems (a breakaway coupling with shut-off valves on either side) should be incorporated into the installation to minimize potential spills during disconnect.

During transfer operations, remote isolation capability should be provided in case of line rupture or uncoupling of connections. Isolation valves should be located directly upstream and downstream of hoses or articulated arms, and should be fail-closed. For long transfer lines, isolation valves need to be designed to minimize surge pressures on closure. Excess-flow check valves can be used, but they operate only at flows exceeding normal operating rates. Thus, they should be used only in addition to, not in place of, remote isolation valves. It is also good practice to visually inspect and pressure-test the connections with purge gas prior to commencing transfers; therefore, local pressure indication should be provided to permit such testing.

After being placed in service, flexible hoses need to be hydrostatically tested or replaced at least once a year. Articulated arms require a purge gas connection at the apex to free the arm of process material before disconnecting. For low-temperature service, they will also have to have purge gas supplied to the joints to dispel moisture. There should also be positive means of preventing a tank truck from driving away while still connected, such as wheel chocks, physical barriers,[1] or by taking custody of the driver's ignition key.

Means for isolation and containment of releases should be provided at the terminal in accordance with the guidelines presented in Chapter 7.

Unloading racks or spotting locations used for the transfer of flammable and combustible materials should be provided with adequate fire protection. While the exact type and design of the fire protection should be a function of the hazards associated with the material being transferred, water spray systems and fixed dry chemical or foam systems are worthy of consideration. Of primary importance is the compatibility of the extinguishing medium with the HTHM. Water may react with certain materials, producing more toxic by-products. Other HTHMs may produce relatively non-toxic products of combustion, in which case it may be desirable to let a controlled fire burn out.

In the cases of flammable HTHMs, fixed pipe systems with application nozzles should be designed to provide water at a specific discharge density with a particular distribution pattern over the area to be protected. The system should be capable of manual or automatic activation and be connected to a reliable water supply. Even if water is not a suitable extinguishing agent, supplies may be desirable to allow cooling of vessels, piping, or supports. Water run-off control provisions need to be addressed.

Similarly, fixed dry chemical systems having fixed piping and nozzles, a reservoir of dry chemical, and an expellent gas for discharge can be employed. Either type of system can be activated by traditional detection systems using heat or flame detectors, or by manual tripping devices. The latter should be located at a contiguous control room or along a means of egress. Inspection and maintenance of these systems are of critical importance to their reliable operation in case of fire.

5.2 Marine Transfer Facilities

Waterfront facilities engaged in marine transfer operations are primarily regulated by the U.S. Coast Guard. Part 126 of Title 33 of the Code of Federal Regulations (33 CFR Part 126) relates to required safety measures for handling explosives or other dangerous cargoes within or contiguous to waterfront facilities. Parts 154 and 156 include facility environmental protection regulations. Although these latter sections focus on oil pollution, they also contain the basis for the safe handling of other hazardous cargoes. Where the regulations do not address concerns relating to design, construction, or operation of transfer systems, it is best to refer to the advice presented elsewhere in these Guidelines, including many of the recommendations already given for rail and highway vehicle transfer operations.

Some considerations, however, are specific to marine transfer operations. The most important is weather. Before start of transfer operations, a current local weather forecast is necessary. Unexpected high winds can cause failure of the transfer lines by movement of the vessel from the dock. The effects of normal and extreme tidal action should also be considered in designing the extension requirements for flexible lines and loading arms.

5.3 Non-Permanent Storage

Rail cars and tank trailers are sometimes used in place of permanent storage facilities. When a temporary storage container is used over extended periods (weeks or months), serious consideration should be given to the installation of a permanent storage facility. The rail car or truck relief valves vent to atmosphere, whereas relief valves on HTHM permanent storage tanks should be connected to a vent header system. When rail cars or tank trailers are used for non-permanent storage of HTHM, they also require adequate containment systems. Some additional aspects are discussed below.

For high vapor pressure or flashing liquids, one approach to containing small or moderate releases is the use of an enclosure. Enclosures are discussed further in Section 7.2.1d. Fixed-point vapor monitors with alarms in the control room should be installed.[3] Additional information is given in the AIChE-CCPS "Guidelines for Vapor Release Mitigation."

The location of spotting areas for non-permanent storage should be given careful consideration. The use of existing rail spurs or unloading racks near the fence line may not be appropriate, particularly in populated areas. The accidental intrusion of locomotives on the spur line should also be considered and appropriate derailers or barricades installed before connecting to transfer lines. In addition, the routing of transfer piping should be kept away from vehicle traffic lanes and protected from impact.

When non-permanent tanks are used to feed a continuous process, careful attention should be given to the method for switching in a new tank so that on-site inventories are minimized, the potential for overfilling one connected tank from another is prevented, and unwanted interruption in supply is avoided.

References

1. Lees, F. P., "Loss Prevention in the Process Industries," Vol. II, pp. 775, 776; Butterworth, Boston and London, 1980.
2. "Safety Advice for Bulk Chlorine Installations," Health and Safety Executive, VIC, Booklet HS(G)28, 1986.
3. "Guidelines for Vapor Release Mitigation," AIChE-CCPS, New York, 1988.

6

INSTRUMENTATION/CONTROLS AND DETECTION

Instruments and controls are an essential part of the safe design of systems for the handling and storage of HTHMs. They are key elements of systems that are designed to maintain and regulate normal operations and to anticipate the onset of, and eliminate the threat of, conditions that could result in loss of containment. They are also used for the early detection of releases, so that mitigating actions can commence before such releases result in serious episodic events.

6.1 Instrumentation, Controls, and Electrical Systems

6.1.1 General Considerations

Typically, in the design of instrumentation and controls for non-hazardous processes, it is not normal practice to deal with double- and triple-jeopardy (i.e., multiple failure) events. This can be justified on the basis that the event is of low probability and that the consequences are not life-threatening. When HTHMs are involved, however, this philosophy is no longer valid because of the potential serious consequences of even low-probability events.

The basic approach to safe instrumentation and controls for hazardous material handling and storage should be to direct the system to the safest operating level relative to employee and/or public well-being, whenever any possible emergency condition is detected, including power loss. Such an approach, if not properly implemented, can create problems from an operational standpoint. Therefore, an important concept of instrumentation and control system safety addresses the task of reducing spurious shutdowns, while maintaining an adequate level of certainty that a safe state will result if a real emergency does occur.

Every reasonable effort should be made to maximize the effectiveness of automatic control systems for preventing an accidental release. All systems and instrumentation should be of the "fail-safe" type. Systems that must be operable under emergency conditions should be designed considering the potential requirements for remote operation and/or manual override.[1] An uninterruptible power supply or backup generation system should be provided for control systems that are critical to safe recovery in emergency situations and in situations where manual operation would be impractical. Instruments should be made of materials capable of withstanding emergency environments (if they have to function under such conditions), as well as the corrosive or erosive conditions to which they are subjected. Control rooms that will be occupied should be protected from intrusion of flammable and toxic vapor as well as fire and explosion hazards.

6.1.2 Measurement Instruments

A. PRESSURE

For local measurement of pressure, diaphragm pressure gauges are preferred. Every effort should be made to minimize the use of direct-connected pressure gauges (Bourdon-tube type). When used, these gauges must be installed with shut-off cocks or valves. Flow limiters (e.g., restriction orifices) may also be used to limit flow in case of loss of integrity, provided that the hazards associated with a small release have been properly addressed. In corrosive service (e.g., bromine, chlorine, hydrogen fluoride), diaphragms and Bourdon tubes should be constructed of appropriate chemical-resistant materials. Where the use of direct-connected pressure gauges is unavoidable, inert liquid filling can be used to combat corrosion, provided the fluid is compatible with the toxic material. The pressure range for each type of pressure-measurement device should be about twice the operating pressure.[2] The materials of construction for wetted parts of transmitters should also be satisfactory for use with the material being handled.

B. LEVEL

An accurate indication of liquid level in containers is critical to the avoidance of potentially hazardous conditions in storage vessels containing HTHMs. Redundant level indicators with different measurement principles (e.g., buoyancy, pressure, capacitance) should be specified with indications and alarms reported in the control room. Reverse-indicating devices (e.g., float systems with indication as distance from container top) should be avoided to eliminate the chance of human error in interpretation.

Capacitance-type level sensors are often employed because they require little maintenance and are highly reliable; however, they should not be used where conductive or high dielectric constant coatings are a possibility. They are also not satisfactory for fluids near their critical point, and may not be suitable where material composition changes widely. Strain gauges or load switches measuring weight should also be considered as a means of indicating level or inventory, since they are very accurate and do not require penetration of the primary containment vessel. If local level gauges are required to operate equipment effectively, a non-glass, magnetically cursored type is preferred. Level alarms and shutdown functions are extremely important in limiting hazards. These are discussed in Section 6.1.3.

C. FLOW

To the extent practicable, non-intrusive or low mechanical-action flowmeters, such as magnetic or vortex-shedding types, are preferred. Orifice plates are particularly easy to use, accurate and reliable (except in non-clean streams), but their range is limited and their use introduces an additional flange and other threaded connections in the piping. The use of rotameters with glass tubes is not appropriate and should be avoided.[3] For small flow measurement and control, metering pumps can be employed.

D. TEMPERATURE

For accurate temperature measurement, intrusion into the fluid is usually unavoidable. Temperature-measuring elements should usually be protected by thermowells; however, thermowells will slow down the measurement response time. Where thermowells are exposed to HTHMs, they should comply with the same materials guidelines as for the equipment in which they are installed to reduce failure from erosion/corrosion. (Refer to Chapter 4 of this document for material and construction guidelines for tanks, vessels, and piping.) To avoid fatigue failures of thermowells in high-velocity gas streams, the manufacturer's recommended insertion length should not be exceeded. Differential sensors are preferred for differential temperature measurements, rather than relying on the difference between absolute measurements. Installation of dual elements should be considered in services where sensor replacement is difficult. In addition, separate elements should be used where redundancy is required.

E. CALIBRATION

The accurate calibration of instruments is of utmost importance for hazard identification and control systems. Instruments and control devices that are critical to safety should be identified and appropriately classified, and this class of instruments should be given priority maintenance. Calibration of instruments should be tied in with a thorough preventive maintenance program (see Chapter 8), and should be done periodically to ensure proper instrument operation. Calibrations are usually best done according to the manufacturer's specifications and, where possible, should be traceable to U.S. National Bureau of Standards reference measurements. In some instances (e.g., on-line analyzers, pH meters, toxic and flammable gas detectors), special maintenance programs are advisable.

6.1.3 Control Systems and Interlocks

Control systems serve several functions that are related to operational safety, including information collection, normal control, and fault administration. These place considerable importance on control systems for HTHMs. The fact that instrument failures do occur has to be taken into account in designing control systems. Consequently, the reliability of instrumentation that is critical to hazard identification and control should be carefully assessed, using techniques such as hazard and operability studies (HAZOPS), failure modes and effects analysis (FMEA), and fault trees. Additional integrity should be provided when the probability of failure upon demand is high, and integrity can be accomplished by redundancy and diversity of components. For example, where system redundancy is needed for safety, the control system critical instrumentation should be supplemented by a completely independent safety back-up system which preferably should incorporate different principles of measurement (diversity). In other situations, complete parallel control systems may be used.

A. OVERPRESSURE CONTROL

The approach to overpressure control in storage systems depends, to a large extent, on the overall design concept, which can be based upon pressure relief or pressure containment. On HTHM systems with small volume, such as a short section of liquid-filled pipe, either relief to a safe location or pressure containment can be used. An example of pressure containment is the use of expansion chambers on a liquid chlorine transfer line to capture the discharge from pressure/thermal relief rupture disks, or the use of a heated expansion chamber that is connected to the pipe. The design of storage tanks and most pressure vessels is based upon relieving the pressure during the design basis incident. The selection of the specific design incident requires careful analysis of all potential causes of overpressure, including both external (fire) and internal (reaction) energy sources (see Chapter 2).

Since the majority of storage tanks utilize pressure relief to prevent high-pressure failure, pressure relief valves and/or rupture disks are necessary to protect against overpressure resulting from process upsets or external fire. If the tank can experience vacuum and is not designed to withstand vacuum, it should be protected with a vacuum breaking system as well. However, where practical, it is preferable to design tanks and vessels for full vacuum rather than rely on vacuum relief systems.

Pressure vessels and vessel jackets should be fitted with adequate pressure relief. Sizing of relief systems should conform to procedures and standards appropriate in the industry such as the API's Recommended Practices 520/521.[4,5] The Design Institute for Emergency Relief Systems (DIERS), under AIChE sponsorship, has also been active in developing techniques for sizing relief systems,[6,7] including those with potential for runaway reactions.[8] Containment and disposal systems should also be designed according to recognized guidelines (see Chapter 7). Where possible, a pressure control or depressurization system should be used along with a pressure relief system. This will help reduce the frequency of releases of HTHMs through the pressure relief system.[9] The adequacy of a pressure or vacuum relief system should be reevaluated when a vessel is used to handle more throughput or a different material from that for which it was originally designed.

To ensure availability on demand, safety valves should be tested periodically and adequately maintained. Therefore, on continuously operating systems and storage vessels that cannot practically be taken out of service, dual pressure relief devices should be provided, with each sized for the maximum contingency. To permit safe removal of either relief device for maintenance and testing, the most inherently safe system utilizes an upstream three-way valve, so that one relief device is always in service on the protected equipment, regardless of the position of the block valve. When the discharge from parallel safety valves enters a common header, a similar three-way valve is provided at the discharge and should be interconnected to the inlet valve so that a direct route to the header is always open. Some codes permit installation of full-area block valves, either upstream or downstream of relief devices, to facilitate removal of such devices for inspection and repair purposes, providing the relief device cannot be readily rendered inoperative.[10] Where such block valves exist, effective administrative locking and tagging procedures must be practiced to ensure that locked or sealed valves are in the appropriate position.

It is important that the safety relief device be constructed of materials that are compatible and suitable for the service. For highly corrosive environments, more frequent testing of relief devices is required. Alternatively, rupture disks made of a corrosion-resistant material can be installed underneath the inlet to the pressure safety valve. If rupture disks are employed, they should be of the non-fragmenting type (graphite rupture disks are inappropriate). When rupture disks are employed, it is important that the pressure downstream of the rupture disk not be allowed to build up and affect the differential pressure imposed on the disk. Therefore, a pressure switch that alarms in the control room, or a pressure gauge connected with a car-sealed open block valve, should be installed between the rupture disk and the safety valve to monitor the pressure and to indicate possible premature failure of a rupture disk. In the former case, periodic testing of the pressure switch is important; in the latter, frequent monitoring of the pressure gauge is necessary. Premature rupture of disks is more likely when the normal operating pressure is close to the burst pressure. Therefore, sizing and selecting the set pressure for rupture disks require careful evaluation. Also, rupture disks that could be subjected to vacuum conditions may need mechanical support on the low-pressure side.[11] Some operators have reported difficulties with reverse buckling rupture disks that rely on knife blades for activation.

Pressure relief devices (except those on transport vehicles used as non-permanent storage) that handle a HTHM should discharge to a closed blow-down header that communicates with a suitable toxic destruct system, such as a scrubber or incinerator. Guidelines for designing venting systems are discussed in Chapter 7.

B. LEVEL CONTROL

Level management in storage containers is of critical importance in preventing a liquid release. The primary objective is always to operate within the working level range of the tank. Avoiding overfilling of tanks is important, since an extremely high level may impair the pressure relief system, cause liquid to be discharged through the relief device, or it may fail the tank. An extremely low level can result in loss of pump suction and possible failure of the pump or pump seal.

Since the level in storage tanks is expected to fluctuate, level management requires a reliable level indication. The need for redundancy and diversity in level sensing has already been discussed. Primary level indication and alarm enunciation, at a minimum, should be located where the operator responsible for controlling material transfers is stationed. Indication in a second occupied location is desirable as well.

Given the frequency of transfers in typical storage facilities, the possibility of equipment malfunction and human error has to be anticipated. High- and low-level alarms should be employed to warn the operator that the normal level range is being exceeded. In the case of a high level, an independent high-high level alarm and trip system that automatically stops flow into the tank should be specified. Alarms and shutdowns should be communicated to the control room, or to a location where an operator is assured to be in continuous attendance. When appropriate automatic level detection fail-safe devices are chosen and installed, they have drastically reduced the frequency of spills resulting from the overfilling of containers.[12]

C. FLOW CONTROL

Control or monitoring of flow in handling and storage is done primarily to account for production inventory or customer transfer. However, there are design aspects of flow regulation that have hazard-control implications.

Control valves are a very important element of the control loop. They should be correctly sized and have appropriate rangeability and control characteristics, including the capability to close (open) against maximum possible pressure drop. They should also have fail-safe features, which may include action on loss of power and suitable limits to flow when fully opened. Bellow seals or tightness of closure can be provided where leak prevention is essential.

A flow-limiting orifice may be inserted in the line to limit the rate of release in the event of a failure. This device is effective when the flow rate is low under normal conditions. The advantage is that a larger, heavier walled pipe can be used in a vulnerable environment without increasing potential release rates.

In some instances, excess-flow valves (EFVs) may be legally required for transfer lines attached to vessels containing significant inventories of hazardous materials (e.g., OSHA - 1910 for ammonia tank cars). However, they may convey a false sense of security, because they do not provide any protection below the level of flow at which they close. Care should be taken to ensure that the EFVs are sufficiently sensitive to respond to large leaks as well as full line failures. Particular attention should be given to the sizing of the piping downstream of the EFVs to ensure that it does not limit the flow below the setpoint of the valve.

D. TEMPERATURE CONTROL

For ambient temperature storage, tank temperature is monitored, but not normally controlled, per se. Temperature indication, in combination with pressure controls, is certainly desirable for identifying one of the potential causes of high pressure. It can also indicate that the tank contents are approaching a hazardous region (e.g., onset of reaction). Also it can indicate thermal stratification in the tank, which sometimes can lead to rapid mixing and vapor release. This phenomenon (sometimes called "rollover") has been occasionally observed in ammonia tanks containing more than 5000 ppm of water.

For refrigerated or low-temperature storage at atmospheric pressure, temperature control may be maintained through boil-off. It can be affected by composition change (multi-component) or transfer of warm product. For pressurized and refrigerated storage, temperature is controlled through the application of external refrigeration, which is usually activated by an increase in tank pressure (preferably with an automatic start sequence). In the case of low-temperature storage tanks, it is recommended that a high-temperature alarm be placed in the transfer lines from the process units to provide warning in the event that excessively warm material is sent to storage.

E. ALARMS AND EMERGENCY SHUTDOWN SYSTEMS

Alarms should function as early warning devices that anticipate the onset of a potential hazardous condition. Alarm setpoints should be established so that enunciation of an alarm indicates that conditions are outside the normal operating range, and action is required by operators to return to normal conditions. Clearly, operator response time should be considered in determining the proper setpoint. The alarm system should operate acording to a properly thought-out logic, which relates the variables alarmed; the number, types, and degrees of alarm; and the alarm displays and priorities to factors such as instrument failure and operator confidence, the distinction between alarms and status information, and the information load on the operator.[3] Alarms that are essential to safety should be identified and classified separately from process alarms, and provided with a different type of enunciator or indicator. Alarms that are no longer needed or meaningful should be deactivated and removed from the system.

When a variable is critical for process safety, a single sensor alone should not be used for an alarm, trip and control function. Safety considerations require a second, independent sensor system. For example, a storage tank containing HTHMs should have a high-level alarm that is independent of the operating level indicators/controllers. The reliance that operators have on interlock systems and the effects this may have on attention to emergency alarms should also be evaluated.

When dealing with episodic releases, it is desirable to have backup systems to deal with situations that have progressed to the point that normal response is inadequate. Emergency shutdown systems are employed to deal with such a situation and should be designed for quick activation and secure containment of the hazard, such that loss of containment is avoided or at least minimized. An example is the high-high level shutdown of a transfer line to avoid the hazards associated with overfilling storage tanks. In the design of emergency shutdown systems, human error should be taken into account.[3] Therefore, in addition to remote shutoff capability, use of automatic shutdown systems that do not require human intervention are appropriate, depending on the seriousness of the consequence. For example, the emergency depressurization system on a vessel containing a reactive chemical should not require operator activation. Proper sequencing of shutdowns must be considered during design of the shutdown system to avoid unnecessary process upsets or equipment damage.

F. COMPUTER CONTROL SYSTEMS

Computer control system functions that relate to HTHM storage and handling include

- self-testing of the computer,
- detection of sensor/operator/valve failure, and
- material and energy balances to determine the "reasonableness" of measurements.

All current control computers test data transmission and internal operations by (a) using a watchdog timer that has to be reinitiated periodically; (b) running a test problem that exercises memory access, arithmetic operations, and analog/digital conversions; and (c) testing transmitted data by parity bits and associated security code. Standard methods of detecting failed sensors/operators/valves include (a) loss of transmitted signal (which is why live zero signals are preferred); (b) change in resistance of the sensor, such as a thermocouple; (c) valve position feedback, or limit switches; and (d) periodic cycling of analyzers such as chromatographs, using a standard fluid substituted for the process fluid.

A separate computer that accesses the control computer's data base is often used to make material and energy balances and to test for "reasonableness." One example is a mass balance cross-check on the quantity unloaded from a storage tank against the weight charged to a batch reactor.

Monitoring for leaks in transfer pipelines by a computer control system is difficult for a variety of reasons. A substantial leak may be only a small fraction of the total amount of fluid being conducted through the pipe or pipeline configuration, and perhaps inaccessability would make point fluid detectors impractical. In addition, the dynamics (particularly deadtime) would make a material balance model very large and time-consuming to solve. While this is being done on large liquid pipelines, it has had limited success within the plant. Finally, acoustic detectors that listen for the sound of a sudden rupture have been successful only for sudden, large leaks. The problem is one of frequency spectrum analysis to identify certain sonic "fingerprints."

One common practice is to implement emergency shutdown (ESD) logic in a computer that is separate from the control computer, with (as mentioned previously) separate sensors and operators/valves. To obtain a high level of reliability, computerized trip systems should be provided with back-up systems. One approach is to have a separate, hardwired alarm/trip system operating in parallel with the computer; another method is to provide a duplicate computer. In either case, the power supply to the ESD system must be maintained during all emergencies. Frequently, programmable logic controllers (PLCs) have been selected for the ESD function since (a) most emergency situations are identified as "trips," or discrete switch openings, and (b) the PLCs have been more thoroughly "hardened" and thus are not subject to maloperation if an air conditioner/dehumidifier fails, as has been the case with control computers.

Process control computer systems with built-in triple redundancy have been available for more than 5 years, but have had limited commercial success, perhaps because of the extremely high reliability of today's control computers. Personal computers (PCs) are not suitable for process control or ESD implementation in high-hazard situations because they are not sufficiently rugged for use in industrial environments.

6.1.4 Reactive Chemical Storage

Special precautions are required for storage of HTHMs that are also reactive (i.e., spontaneous decomposition, polymerization, etc.). The principal concern is the exothermic heat generation and associated rapid pressure increase in the container. Storage temperature is the critical measured variable since it is a signal that

(a) reaction is taking place and the tank contents are warming up, or
(b) thermal energy is being transferred to the tank and the contents are approaching a critical temperature where reaction rates may be uncontrollable.

Tanks for the storage of reactive materials should be provided with multiple-temperature indicators and alarms. For large tanks subject to stratifiction, temperature indication at various levels in the tank is appropriate. The tank should also have a high-high temperature interlock system that automatically activates an emergency system, such as supplying cooling or quench (dilution) to the tank contents.

Several methods are employed for controlling temperature of storage containers. Both external and internal arrangements can be considered depending on the potential hazards involved. External methods can be as simple as using reflective paint or insulation, or they can involve the use of water films or jackets. Internal systems can consist of

(a) internal coils for circulating coolant,
(b) reflux condensers that return the condensate to the tank,
(c) quench systems that inject an inert liquid solvent or diluent to lower reactant concentration and temperature, and
(d) short stop systems.

Adequate mixing is important to the efficacy of all of these systems.

Another important consideration with reactive materials is control of contamination in storage. Contaminants such as oxygen, iron rust, water, acids, and bases can often initiate vigorous reactions. Another source of cross-contamination is the backflow of inerting gas from other production units. This can be minimized by use of non-return valves, and making sure that the inert gas used for reactive storage vessels is supplied from the first lateral off the main gas supply header. In particularly hazardous situations, continuous monitoring of the inert gas supply may be appropriate to avoid contamination.

Another potential source of contamination is the unloading and transfer of off-specification or inappropriate materials. Quality assurance procedures and administrative controls are quite often the most effective defense.

6.1.5 Electrical Wiring and Safety Barriers

All wiring and electrical equipment should be installed in accordance with the National Electric Code (NFPA 70). Electrical equipment exposed to flammable HTHMs in areas that should be classified according to Article 500 of NFPA 70 should meet the requirements of that article. Electrical instrumentation may be intrinsically safe in accordance with the National Fire Protection Association's Standard for Intrinsically Safe Apparatus and Associated Apparatus for Use in Class I, II, and III, Division 1 Hazardous Locations (NFPA 493) and the Instrument Society of America's Recommended Practice 12.6 (ISA RP 12.6), purged in accordance with NFPA's Standard for Purged and Pressurized Enclosures for Electrical Equipment in Hazardous Locations (NFPA 496), or treated as other electrical equipment.

In addition, conduit containing electrical wiring, originating/terminating or passing through areas handling HTHMs, and penetrating control houses or other buildings of human occupancy, should be sealed in accordance with the requirements of Article 500 of NFPA 70, even if flammable materials are not of concern. This is to prevent the transmission of nonflammable toxic vapors into locations that may be "safe havens" in a toxic release emergency and may have to remain clear for proper response actions by operators.

6.2 Release Detection

This section deals with the detection of toxic materials once a release has occurred. A more comprehensive treatment of release detection is contained in the AIChE-CCPS's "Guidelines for Vapor Release Mitigation." (Appendix C lists the contents of those Guidelines.) Early detection of releases is essential for identification of appropriate response actions.

6.2.1 Monitors/Detectors

Fixed-point monitors can be useful for detecting both major releases and small leaks of toxic materials so that the proper response action can be initiated at the earliest time. To provide the highest confidence that a release of a toxic material will be detected, care must be taken in placing the detectors, because their ability to detect may be affected by wind. Fixed-point monitors are particularly useful in the immediate vicinity of rotating equipment, such as transfer pumps that are prone to seal leaks. Other areas where fixed-point monitors should be considered are loading/unloading manifolds, piping manifolds at bulk storage tanks, inside buildings and structures that contain HTHMs, and at areas around remote storage tanks. Location of the sensing head relative to grade needs careful consideration, taking into account the physical state of the released material as it disperses. For example, the cold gas released from a cryogenic storage tank may behave like a dense vapor, even though under ambient conditions it would be a buoyant gas. More information on the positioning of fixed-point sensors is given in Reference 9.

Fixed-point monitors also require frequent calibration and testing. This calibration should be done with an appropriate reference gas where the species of interest is diluted to a level consistent with the expected monitoring range of the sensors. The calibration should be performed at two points over the concentration range of interest.

6.2.2 Monitoring of Vents

Continuous vents, such as scrubber and incinerator stacks, and clean-up adsorbent bed vents should be provided with detectors to continuously monitor for breakthrough of toxic substances. Redundant sensors should be installed to monitor flare tips to ensure that the pilot burner is always lit. Since these detectors are the last line of defense, they should have a high degree of reliability and integrity. Vent detectors should be provided with an alarm function set at a level that allows adequate response before the hazard of breakthrough becomes significant. Since these monitors will require frequent testing and maintenance to ensure operability, reasonable access should be provided. In many applications, redundant detectors should be installed to provide the required reliability. Frequent testing of the detectors for threshold sensitivity and operability should be performed.

References

1. Draft Summary Report: "Technical Support to the South Coast Air Quality Management District Toxic Chemical Accidental Air Releases," Radian Corporation, Austin, Texas, November 12, 1986, DCN 86-203-024-84-13.
2. "Standards and Practices for Instrumentation," 6th ed., 1980, Edited by L. M. Ferson, Instrument Society of America.
3. Lees, F. P., "Loss Prevention in the Process Industries," Vol. I. Butterworth, Boston and London, 1980.
4. "Guide for Pressure Relieving and Depressuring System," API Recommended Practice 521, 2nd ed., p. 44, American Petroleum Institute, September 1982.
5. "Recommended Practice for Design and Installation of Pressure Relieving Systems in Refineries," API RP 520, American Petroleum Institute, 1976.
6. Fauske, H. K., "Emergency Relief System (ERS) Design," Chem. Eng. Prog., August 1985.
7. Leung, J. C., "Simplified Vent Sizing Equations for Emergency Relief Requirements in Reactors and Storage Vessels," AIChE Journal, Vol. 32, No. 10, Oct. 1986.

8. Fauske, H. K., "Generalized Vent Sizing Monogram for Runaway Chemical Reactions," Plant/Operations Progress, Vol. 3, No. 4, Oct. 1984.
9. "Guidelines for Vapor Release Mitigation." AIChE-CCPS, New York, 1988.
10. "Boiler and Pressure Vessel Code," Section VIII, "Unfired Pressure Vessels," Paragraph UG-125(d) and Appendix M-5 and M-6, American Society of Mechanical Engineers, 1986.
11. Draft: Prevention Reference Manual - Chemical Specific, Volume B: Chlorine, Radian Corporation, Austin, Texas, November 1986, DCN 203-024-84-16.
12. Heard, David B., "Fail-Safe Devices for the Prevention of Hazardous Materials," U.S. Environmental Protection Agency, EPA 600/S2-84-189.

7

ISOLATION AND CONTAINMENT

The section on isolation describes use of valves and other devices for minimizing releases. The section on containment deals with secondary containment and destruct systems as countermeasures, in minimizing the impact of releases and spills that threaten the environment.

In the event of a system leakage, isolation is one method for limiting the quantity and/or rate of material released, which in effect limits the consequences as well. Containment refers to methods of preventing the release of materials to the environment from the equipment or facilities in which they are to be handled or stored. The term release includes spills, escapes, overflows, discharges, ventings, and leaks. In this context, the definition of containment has been broadened to include confinement of releases to areas close to the origin and away from people and other exposures so that the consequences of the release are minimized.

Primary containment includes the principal vessels, pumps, and piping used in the normal handling and transfer of HTHMs. Secondary containment includes back-up containers, enclosures, dikes and other systems which can safely contain a HTHM in the event of loss of primary containment. Examples include the outer wall of double-walled tanks, and headers for relief systems.

Disposal or treatment refers to the absorption, destruction, etc., of toxic materials. This is discussed more extensively in the AIChE-CCPS "Guidelines for Vapor Release Mitigation."

7.1 Isolation

The general approach to minimizing equipment leakage is to design it so that the flow of fluid to the leak point can be reduced promptly and then stopped or limited. Naturally if the inventory in equipment is very small, isolation systems are not needed. Isolation is achieved by use of valves and flow limiters.

7.1.1 Isolation Valves

For HTHMs, significant releases can be as small as a few pounds, depending on the material, distance to the public, etc. Isolation valves are used to limit releases to amounts that will not have community impacts. Isolation valves can be separate, remotely operated block valves (RBVs), or control valves with separate shut-off action. When separate RBVs are used, they have to be tested frequently to ensure they are always available upon demand. They should be installed as closely as possible to the vessel.[1]

Isolation valves should be installed on HTHM storage tanks. They are normally installed on bottom connections to liquid-filled tanks and are often used in addition to excess flow valves that may not give a tight shutoff. However, they should also be considered for top connections on pressurized storage tanks containing high vapor pressure materials and on lines with dip tubes. Long pipelines also may have isolation valves installed to limit the amount of material (inventory) that might be lost if a break occurred. Isolation valves in or near flammable-fluid service should be of fire-resistant design, or protected from possible fire by water deluge or insulation and fire shields.[2,3] For further discussion on isolation valves, see Section 4. Isolation valves should be remotely operable and of fail-safe design.

7.1.2 Other Devices

Spring-loaded check valves can be used to prevent backflow from storage tanks to unloading stations in the event of a leak in a flex hose or articulated arm joint. However, check valves should not be depended on for complete shutoff; in critical locations they should be backed up with isolation valves.

Remote shutdown of pumps can also reduce the rate and duration of leakage. Remote shutdown of loading arms from a position near the loading racks should be provided in the event of a hose or arm leak. Switches for operating motorized isolation valves and shutting off pumps should be nearby and at a safe distance from points of potential leakage. Two switches separated by at least 30 feet are preferred.[5] Isolation systems involving automatic closure of valves when loading arms and hoses are overextended, or when tank cars are accidentally moved, should be used.

A flow-limiting device may also be inserted in the pipe to limit the rate and quantity of the HTHM that can be released. This is effective when large flows are not required through the line under normal conditions. For example, flow-limiting orifices are often used in instrument piping. Their use allows a larger, heavy-wall line to be employed in a vulnerable situation without increasing the flow of material.

7.2 Containment

7.2.1 Controlled Transfer

A. SECONDARY CONTAINMENT

The loss of primary containment can occur as a result of pressure relief or emergency venting, and loss of integrity due to leaks, cracks, ruptures, etc. Depending on the seriousness of the consequences of a release, backup systems can and should be designed to guard against loss of primary containment, particularly in the case of HTHMs. These backup systems are referred to as secondary containment systems.

Secondary containment systems in this context include closed pressure relief/vent headers, double-walled vessels and piping, dump tanks, and enclosures. The term secondary containment is sometimes extended to include dikes, curbs, and trenches used to contain spills in the immediate vicinity of a release. These storage area features are discussed in AIChE-CCPS "Guidelines for Vapor Release Mitigation."

The material collected by secondary containment systems must ultimately be safely removed. Consequently, disposal or destruct systems for HTHMs are also addressed later in this section. Provision must also be made for removal of rain water from open secondary containment systems in a manner that does not compromise the integrity of the containment. For example, an "over-the-top" pump-out may be preferred to a drain line and valve penetrating the dike wall below full liquid level.

B. VENT HEADERS[4]

Release prevention (integrity) is the most important aspect of release mitigation and containment, because it is the most effective. Overpressure vent headers or closed blowdown systems should be used to collect the releases from relief valves and emergency depressurizing valves on HTHM storage vessels. Although there are dangers from manifolding releases from various sources that must be addressed, where practical the discharge piping from relief devices should be combined into a common header system, which transfers any relief discharges to a knockout drum or catch tank for trapping liquids before disposal to a scrubber, flare, or other destruct system.

The design of headers requires taking into consideration the system loads, reaction forces, the backpressure limitations, the requirement for special materials, the presence of corrosive materials, the reactivity of all discharged materials, the significant differences in the pressure levels of the equipment connected to the system, and the possibility of abnormally high or low temperatures.[6] Also, different frequencies of maintenance and shutdown and the possibility of air entering headers for flammable gases or vapors may affect the arrangement and routing of headers. When the material involved is both toxic and flammable, the introduction of a sweep gas to prevent backflow of air into the header is good practice. The sweep gas should be non-reactive with the toxic material, with possible choices of nitrogen, carbon dioxide, and natural gas in the case of a flare header. The possibility of liquid discharges into headers may require local knockout drums, special piping arrangements to prevent liquid from contacting the discharge parts of relief valves and rupture disks, and prevention of freezing, solidification, or polymerization.

Depressuring valves may be installed in parallel with relief valves in order to control the pressure below the relief valve set pressure, to avoid operation of relief valves which may not reseat properly, etc. The effect of their use on vent header piping size needs to be considered.

C. DOUBLE-WALLED EQUIPMENT

If the consequences of a release are exceptionally severe, because of large quantities of HTHMs or the proximity of people, double containment should be considered. Large cryogenic ammonia tanks (following designs for liquefied natural gas tanks) are sometimes enclosed in secondary containers, with methods for detecting leaks within the annular space.[1] Sometimes double-walled, single-roof tanks are used to provide secondary containment of liquids with the vapor in the annular space. A high-integrity piping design can include double-walled piping, with the process fluid in the inner pipe and the purge gas in the annular space coupled with gas analysis for detecting leakage (see Chapter 4 for additional discussion on the design of double-walled tanks and piping).

D. ENCLOSURES

Enclosures are containment structures that capture any spilled or vented material from storage or unloading systems, thereby preventing immediate discharge of a HTHM to the environment. The enclosures contain the spilled liquid or gas until it can be transferred to another containment facility, or transferred at a controlled rate to scrubbers or incinerators for destruction.

The use of specially designed enclosures for chlorine and phosgene storage or unloading equipment is practiced at some locations. Opinions vary on the appropriateness of enclosures for HTHM operations.[1] The feasibility of using an enclosure depends partly on the frequency with which personnel must be involved with the equipment. A problem with having enclosures for HTHM is the accumulation of toxic concentrations from smaller leaks within enclosed areas. Ventilation and vent gas treatment systems have to be designed for such conditions. However, if the issue is protecting the community from larger accidental releases, then total enclosure may be appropriate. Enclosures should be equipped with continuous monitoring equipment and alarms. Alarms should sound whenever incipient harmful concentrations are detected. In addition, access to the structure should be limited to authorized personnel who are required to wear breathing apparatus before entering the area.

The enclosure building should be gas-tight and purged by a ventilation system designed to draw in air while the building is being vented to a scrubber. The bottom section of the building, if used for stationary storage containers, should be liquid-tight, and the floor should slope toward a collection sump to retain spilled liquid. Floors sloped toward a liquid collection sump are also used for loading/unloading areas, including vehicles or rail spurs, with a vapor seal across the vehicle entry point. The preferable sump geometry should be narrow and deep to minimize the surface area exposed to the environment. Ideally, the building also should be designed to withstand the potential overpressure generated from the "worst-case" release. When this is not feasible, overpressure venting has to be provided to avoid structural damage. If the building also contains materials that are flammable, reactive, or explosive, these effects must also be considered.

E. DISPOSAL SYSTEMS

As indicated above, secondary containment systems are installed for the temporary collection, or holding, of inventory during an emergency. Secondary containment for design basis incidents involving HTHMs should always be coupled with systems designed for destruction and/or ultimate disposal. Appropriate disposal systems include

- scrubbers,
- passive scrubbers,
- stacks,
- flares,
- catch tanks for vapor-liquid separation,
- incinerators,
- absorbers,
- adsorbers,
- condensers, and
- vacuum trucks (for off-site disposal).

Relevant design information on disposal systems is contained in the AIChE-CCPS' "Guidelines for Vapor Release Mitigation." A listing of relevant topics in those Guidelines are given in Appendix C. Except for relatively low concentrations of gases and vapors, direct venting through stacks is no longer an acceptable practice for HTHM. There are legal and regulatory requirements that limit stack emissions. Further discussion of design practices for stacks is also included.

7.2.3 Release Countermeasures

There are many techniques available for mitigating the effects of releases. These are also covered in detail in AIChE-CCPS' "Guidelines for Vapor Release Mitigation."

These include countermeasures for vapor and gas releases:

- water sprays,
- water curtains,
- steam curtains, and
- air curtains; and

for liquid releases:

- dilution,
- neutralization, and
- covers; and

for flammable releases:

- deliberate ignition, and
- ignition source control.

References

1. Lees, F. P., "Loss Prevention in the Process Industries," Vol I, pp. 772-773, Butterworth, Boston and London, 1980.
2. Arant, J. B., "Fire-Safe Valves - An Overview," Annual Symposium on Inst. for the Process Industries, College Station, Texas, Jan. 21, 1981.
3. Starkweather, L., "The API Fire Test Standard for Soft-Seated Ball Valves," Plant Engineering, 32 (23):115 (Dec. 7, 1978).
4. "Guidelines for Vapor Release Mitigation." AIChE-CCPS, New York, 1988.
5. Kletz, T. A., "Emergency Isolation Valves for Chemical Plants," Chem. Eng. Progress, 71 (9): 73, Sept. 1975; AIChE Loss Prevention, 9: 134, 1975.
6. "Guide for Pressure-Relieving and Depressuring Systems," API Recommended Practice 521, 2nd Ed., p. 44, American Petroleum Institute, Sept. 1982.

8

PREVENTIVE MAINTENANCE AND INSPECTION

Preventive maintenance (PM) consists of any inspection or testing conducted on equipment to detect impending or minor failures and to mitigate their potential before they can develop into more serious failures. A comprehensive preventive maintenance program is vital to maintaining the integrity of a toxic chemical facility. Early detection and repair or replacement can significantly reduce the potential for a catastrophic failure. Some factors that are important in a preventive maintenance program are listed below:

- identification of critical equipment,
- inspection frequency,
- inspection method,
- inspection procedures and training,
- documentation and analysis of data, and
- legal requirements.

Selection of the proper inspection frequency is determined by experience or, for new facilities, from manufacturer's data and by estimating the likelihood of experiencing a catastrophic failure or a minor failure that could develop into a more significant failure.

Use of a standardized PM system helps to ensure that the proper results will be obtained and can be compared with historical records. Documentation and analysis of data will help to uncover chronic problems that could result in acute failures, will help set proper inspection intervals, and will allow the experience gained on maintaining specific equipment to be applied to other similar equipment. This is a vital element of an effective PM program -- one that requires staff with specific qualifications in PM data analysis. Finally, it is important to have detailed procedures written and inspection personnel well trained to help ensure that the method is being applied properly and consistently from one inspection to another and by different inspectors.

Some types of PM may be required by law or by the insurance underwriter, depending on the type of chemical and the service. All applicable regulatory requirements must be reviewed before a PM program is developed. Recommended practices by trade organizations should also be reviewed.

For new facilities, the need to take equipment out of service in order to perform certain PM activities should be considered in the design stage; otherwise, a certain amount of flexibility in early failure detection may be lost, or more importantly, the PM work will not be done.

8.1 Development of a Program

An effective preventive maintenance program is developed over time as site-specific failure rates are determined.[1] To start a program, historical failure rates have to be determined first, and for each piece of equipment an appropriate preventive maintenance strategy must be developed, as shown in Figure 8.1. In selecting which equipment should be given highest priority for development of a PM program, the following factors have to be considered:

- toxicity of material handled,
- quantity of material stored,
- operating conditions,
- corrosivity,
- equipment characteristics,
- failure rate history, and
- condition of equipment.

It may be difficult to obtain relevant historical failure rate data for all pieces of equipment from published sources. Available data have to be compared with site-specific applications. Areas in which the available data will most often differ from the system under review are summarized below:

- service conditions (temperature, pressure, process material),
- material of construction,
- level of maintenance,
- in-service time,
- startup/shutdown frequency, and
- number of demands.

In many data sources, specific details on the items cited above are not available, so the limitations must be considered carefully before the data can be used in developing a new PM program. If possible, site-specific or process-specific data should be

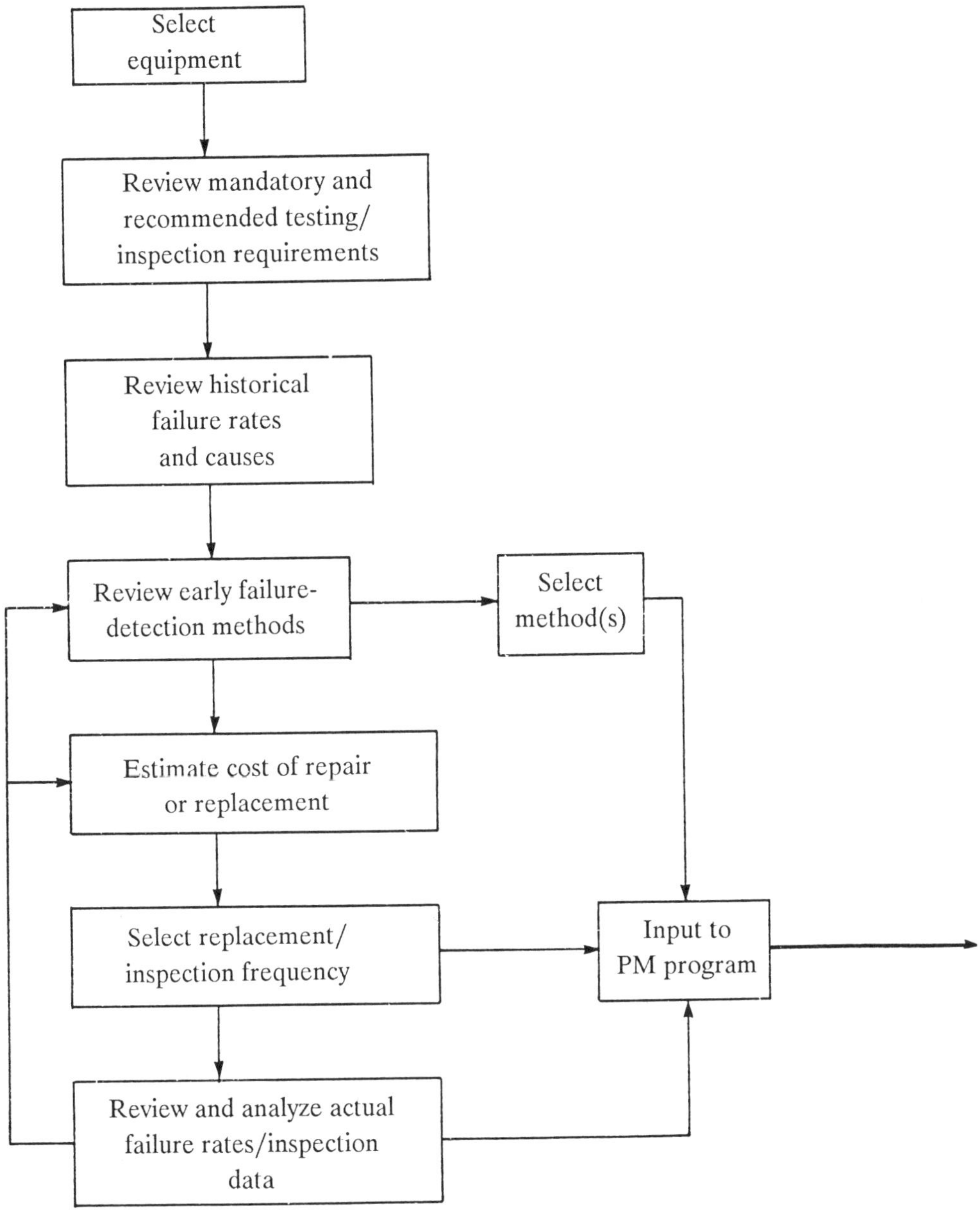

Figure 8.1. *Program Development*

sought. Some testing/inspection of equipment may be mandated by federal or state law, or the insurance underwriter. These requirements must be investigated prior to the selection of methods and frequency. Also, many trade organizations and process licensors have developed recommended testing/inspection schedules for equipment (e.g., NFPA, ASME, API, Chlorine Institute). ASME is currently considering a probability analysis-based approach for determining PM inspection intervals for code applications. These sources should be consulted for more detailed information.

To date, the federal government has not regulated PM, except for pipelines; transportation via air, rail, road or water; and pollution control.

Once the available historical failure rate data have been obtained, the next step is to review the available means of detecting potential failures at an early stage. A review of inspection/testing methods is discussed in Section 8.3. After selecting appropriate methods for inspection/testing, the costs of alternative courses of action must be evaluated. Some courses of action might include (1) test/inspect periodically (e.g., vibration analysis) and take appropriate action, (2) perform PM periodically (e.g., adjust belts) without previous test, and (3) replace periodically (e.g., pump seals, hoses, rupture disks) without PM or prior test. The analysis should consider all the possible impacts of failure. The outcome is a frequency for taking action either in the form of inspection/testing or repair/replacement.

Finally, based on analysis of test/inspection data and actual failure rates experienced, existing preventive maintenance schedules may be changed. Lack of detectable changes may justify extending the interval between testing/inspection programs, while indication of wear or deterioration may require repair or replacement. This feedback may change the approach and/or frequency of maintenance in the future. Any PM program should be reviewed periodically to evaluate new testing/inspection methods, new equipment design, and whether training and skill levels are adequate.

8.2 Inspection/Testing Methods

The simplest form of preventive maintenance is inspection. Inspection focuses on detection of small defects or changes in operating conditions before they result in catastrophic failures. Inspection methods can range from visual to use of sophisticated instruments (see Table 8.1).

Some potential problems that may be revealed by an inspection include[2]

- leaks,
- internal or external corrosion,
- surface defects or wear,
- weld defects,
- deposits and debris,
- high-stress situations,
- vibration or unusual sounds, and
- loss of process or utility flow.

TABLE 8.1
Inspection/Monitoring Methods

---

1. Sensory
Visual: leaks, vibration, surface defects, deposits
Audible: vibration, friction
Olfactory: leaks
Touch: vibration, temperature

2. Instruments
Ultrasonic (wall thickness)
Infrared (temperature)
Radiograph (welds)
Cathodic protection (underground piping)
Vibration (rotating equipment)
Gas detectors (leaks)
Spark (glass-lined vessels)
Acoustic emission (material defects)
Magnetic-particle methods (material defects)
Eddy current methods (ferromagnetic material defects)
Dye-penetrant methods (cracks)
Torque (rotating equipment)
Speed (rotating equipment)
Position
Strain (cracks)

3. Other
Coupons (corrosion)
Analysis of deposits or debris in equipment or lubricating fluids (wear or corrosion)

Operating experience may reveal other potential problems with specific equipment that can be mitigated by routine inspection.

Monitoring for corrosion is important because a significant reduction in the strength of the vessel or line may occur before a leak develops, thereby allowing time to detect and repair this condition. Surface or wear defects are typically found in the form of cracks, pits, or erosion. Avoiding surface defects is significant where there is potential for fatigue. Detection of cracks may be done by acoustic emission testing or by magnetic-particle or dye-penetrant techniques. Deposits can cause serious safety hazards if vent or instrument lines become plugged, or valves cannot seat properly. Certain deposits can also result in unexpected reactions with process fluids or materials of construction (e.g., carbonization).

Flow or pressure-drop measurements can help detect buildup of deposits. Ferrography can be used to provide valuable information about the condition of rotating equipment from analysis of wear particles in the lubricating oil.[3] Excessive vibration can result in catastrophic line failures, but frequently can be readily detected by visual observation. High vibration frequently occurs in poorly supported lines. Visual inspection for external defects can be effective if done in a systematic fashion. For example, it is good practice to walk along pipelines to check for leaks, and to inspect flexible hoses prior to use. Leaks from underground pipelines can sometimes be detected visually by changes in adjacent vegetation or ground surface. This form of inspection can be performed using a simple checklist that the operator fills out during his rounds. Inspection for leaks may be part of a checklist completed each shift, while inspection of hoses might be part of a checklist to be completed prior to each use.

Other forms of sensory inspection may also be included on checklists. For example, the sound produced by a piece of equipment may indicate a problem, particularly for rotating equipment, and the vibration level may also warn of potential problems. Reporting of such unusual conditions should be required so that a more detailed follow-up investigation can be conducted.

Other forms of inspection may require use of special tools. One common device is an ultrasonic meter for monitoring the change in metal thickness in vessels and pipelines caused by corrosion or erosion and for detecting weld defects. Other devices may be installed permanently, such as vibration monitors on rotating equipment. Monitoring instruments themselves should be inspected on a regular basis to ensure proper operation. This may also be included as part of an operator's daily checklist, for example, to verify that monitoring devices have an adequate utility supply (electrical power, cooling water, steam) and proper process flow. All instruments should be recalibrated on a regular basis to help ensure accurate readings (see Chapter 6). The frequency of such calibration depends on the type of instrument and the service, and may range from daily calibration for some analyzers to annual calibration or inspection for other instruments. Further discussion of frequency inspection/testing is given in Section 8.3.

Testing equipment for defects can be time-consuming and expensive. In many instances, however, testing can avoid catastrophic consequences. Often it is required by law or at least recommended by trade organizations. Testing procedures are generally divided into two classes: destructive and non-destructive. An example of the former is a Charpy impact test for low-temperature service. Destructive testing is usually required only in construction or in analysis following a failure. Non-destructive testing methods are preferred for PM. Non-destructive test methods (see Table 8.2) include such procedures as checking relief pressure for relief valves and actuation points for alarms and interlocks, including confirmation that equipment responds in the proper fashion (e.g., that a valve closes). However, even non-destructive testing can cause wear, so these effects and wear caused by preparing equipment for testing must be considered.

Acoustic emission testing of vessels and piping is becoming more accepted as a major non-destructive test method, particularly for HTHM facilities.[4] The method can be used to assess the structural integrity of equipment under operating loads. It can also be used to detect certain types of defects not detectable by other methods, such as

- early-stage stress-corrosion cracking,
- deficiencies in FRP equipment, and
- oxygen and nitrogen contamination of zirconium and titanium welds.

In some cases, testing requires shutdown of the equipment, either to conduct the test or as a result of the test. Therefore, the required testing frequency of equipment must be carefully considered during the equipment design stage.

TABLE 8.2
Some Non-Destructive Testing Methods

- Acoustic emission (lines and vessels)
- Hydrotesting (lines and vessels)
- Ultrasonic testing (lines and vessels)
- Leak testing (lines and vessels)
- Dye-penetrant testing (lines and vessels)
- Relief valves (relief pressure)
- Alarm, interlock setpoints (actuation point)
- Valve operation/emergency shutdowns (confirmation of actual movement to a fail-safe position)

8.3 Frequency

Preventive maintenance has been defined as "the system of determining the probable frequency of failure of a device or system, and inspecting, repairing, or replacing it before it fails."[5]

Determining the optimal preventive maintenance frequency is the most difficult part of any PM program. Even when frequencies are mandated by law, it is sometimes desirable to increase the frequency. As discussed earlier, historical failure rates can provide valuable information, but they must be used realizing that different service conditions can have a significant impact on failure rates.

A target failure rate must be developed for each piece of equipment and then compared to the rates estimated from historical data. Those failures that by themselves could result in potential injury (e.g., vessel or piping failure) will require lower target failure rates than failures that must be combined with other simultaneous failures to result in a catastrophic event. (Acceptable risk is discussed in Chapter 2.) In the latter case, a combination of failures would be required, so other failure rates would have to be considered in determining the target failure rate for a piece of equipment.

Given a failure rate, deciding when testing, inspection, repair, or replacement should be conducted within the time between failures is dependent on a number of factors, the most important being the potential consequences of failure. Thus, the more serious the potential risk, the shorter the PM interval. Other factors to be considered are as follows[1]

- age of equipment,
- severity of service (extremes or cycling of temperatures and pressures, or high corrosion rates),
- service hours (batch operation),

- startup/shutdown cycles,
- unusual ambient conditions (high humidity, extreme temperatures, corrosive environment), and
- potential for natural events (hurricane, tornado, flood).

The impact of a regular testing/inspection program can be easily demonstrated. For a component that is regularly tested and has a negligible repair/replacement time, the average unavailability is[6]:

$$\epsilon = \tfrac{1}{2} \lambda \tau \tag{8-1}$$

where ϵ = unavailability,

λ = failure rate (hr^{-1}), and

τ = test interval (hr).

For a component that is never tested:

$$\epsilon = 1 - e^{-\lambda t} \tag{8-2}$$

where t = service time (hr)

From Equation 8-2, if $t = 10^5$ hr and $\lambda = 10^{-5}$/hr, then $\epsilon = 0.632$, that is, on the average, the component is unavailable 63.2% of the time. However, if $\tau = 1000$ hr (Equation 8-1), then $\epsilon = 0.005$, and the component is unavailable, on average, only 0.5% of the time.

The above simplified analysis suggests that more frequent testing reduces the chance of the equipment being unavailable at a given time because of an undetected failure. This conclusion is correct as long as the time required to test the device is small compared to the service time between tests. It further implies that if testing time is long, or the equipment cannot be taken out of service at the required testing frequency, redundant systems may be necessary and would have to be considered in designing new facilities (see Chapter 4) as well as modifying existing facilities. Table 8.3 lists selected typical testing/inspection frequencies recommended or employed for toxic facilities. These frequencies can vary significantly, depending on site-specific conditions, and should be used only as a guide, not as a specific requirement. If a failure occurs, a thorough investigation should be conducted to determine the cause and define appropriate mitigating measures that should be taken. They might include changes in design or testing/inspection frequency.[7]

TABLE 8.3

SELECTED PM TESTING/INSPECTION FREQUENCIES[7,8]

TYPICAL FREQUENCY	*METHOD*	*EQUIPMENT*	*AREA OF CONCERN*
• Daily	Sensory Inspection	All	Leaks, vibration, external corrosion, wear, insulation
• Daily	Inspection	Monitoring instruments	Utility and process flows
• Weekly	Calibration	Analyzers, detectors	Fouling, deactivation
• 2 months	Inspection	Cathodic protection rectifiers	Underground piping
• 6 months	Inspection	Vessels and lines	Corrosion coupons, external corrosion, insulation
• Annual	Inspection	Rupture disks, flame arresters,	Debris, proper operation conservation vents, and pressure-control devices
	Testing	Grounding systems	Sparks
	Testing	Relief valves, alarms, interlocks	Actuation setpoints and proper activation of control equipment
	Testing	Vessels and lines	Wall thickness, glass lining, acoustic emission
	Testing	Shutoff/isolation valves	Proper operation
• 5 years	Testing	Hydrotest* and/or internal inspection	All vessels
	Testing	Hydrotest*	Underground piping
• Prior to use	Inspection	Hoses, flexible connections	Wear, damage, leaks

*Hydrotesting is not recommended for equipment containing materials that are corrosive in contact with water (e.g., chlorine).

8.4 Procedures and Controls

Each PM activity should have a detailed written and approved procedure to help ensure the quality of the work. It should also provide a document that can be reviewed and updated periodically. The procedure should specify the intent of the activity, the frequency, the required skills, any special tools or equipment required, the required status of the equipment to be tested and of adjacent equipment during the test, and a detailed description of the test procedure. A long procedure should be condensed into a checklist that the maintenance personnel can use in the field. Suitable training must be provided for maintenance personnel in the use of special instruments and test procedures, and operators should be trained in sensory inspection techniques.

Whenever a change in the procedure is required, the procedure should be updated, reviewed, approved, and then redistributed. Outdated material should be destroyed. Relevant personnel should be made aware of the change and retrained accordingly. All procedures should be reviewed periodically, regardless of whether specific changes are necessary.

Positive material identification (PMI) procedures form a special part of any PM program. These procedures are usually developed for plants under construction, but are equally important during ongoing plant maintenance. Many materials, particularly metals, cannot be positively identified by visual inspection. Vendors sometime ship the wrong material, so that without a PMI program improper materials of construction could be installed, risking catastrophic failure.[8] The key items in a PMI program are as follows:

- Purchase materials from reliable vendors willing to supply required materials test reports.
- Accept vendor shipments only if materials have been properly tagged.
- Conduct additional on-site testing/inspection, as appropriate.[8]
- Establish procedures to ensure dispensing of proper materials from plant stores.
- Maintain plant files to enable type of material to be traced.

PMI programs are becoming more common in the United States as companies handling toxic chemicals are realizing the need for such systems and developing programs for their internal use. In much of Europe, particularly the Netherlands, comprehensive PMI programs are mandated by law, facilities are subject to inspection by regulatory agencies, and penalties for non-compliance are severe.

8.5 Documentation and Analysis of Data

Once PM methods and frequencies have been selected, it is important to ensure that the required activities are conducted on schedule. A system must be developed whereby, at the specific time, the PM activity required is performed. A computer is well equipped to handle this type of activity, and many software companies offer programs for just such applications.[7,9]

Upon completion of the work, the findings should be documented to maintain an historical record. This information should be filed in the equipment maintenance record. The file should include a copy of the work order or other document that outlines the work completed and the findings of the test. For PM regulated by law, documentation provides proof of compliance and is often required.

The system must also include an analysis of the test results and a comparison with prior test readings or historical data. This is particularly important for certain tests, such as wall-thickness checks and instrument calibration. The analysis must be conducted by qualified individuals who should be specified by title in the procedure. This analysis is necessary to define trends in the data that may require long-range planning to repair or replace the equipment. Using actual test/failure data, one can program existing microcomputer data base management systems to predict when detailed inspection or maintenance should be performed.[3] This predictive type of preventive maintenance allows maintenance work to be performed based on the actual condition of equipment rather than according to a generalized schedule.[10,11]

It is also necessary to pinpoint when equipment or materials of construction are not suited for the application, or when the equipment is not being operated properly. For example, an instrument that requires frequent calibration may be giving the operating personnel a false sense of security since quite often it will not be reading accurately. In this fashion, actual data from the PM program can be used to refine and develop the PM program over a period of time. Some PM data may be too technical or detailed for transmittal directly to maintenance or operations. Feedback to these groups should be provided in a form that can be readily understood and evaluated.

Proper documentation and analysis of findings is a critical element of any effective PM program. Adequate time and personnel have to be allocated to this activity to help detect gradual changes in equipment, and to repair or replace equipment before a catastrophic failure occurs.

An effective PM program requires all the above aspects to be addressed in the same level of detail. However, if managed properly, a PM program can improve safety and reduce costs.

References

1. Wyder, C. G., "Preventive Maintenance," from Maintenance Engineering Handbook, 3rd ed., McGraw-Hill, New York, 1977.
2. Lees, F. P. "Loss Prevention in the Process Industries," Butterworth, Boston and London, 1980. This gives data on instrumentation and control devices, such as valves.
3. Madhaven, S., and Sathe, S. Y., "Inspection of Ammonia Plants," Plant/Operations Progress, p. 35, January 1987.

4. Fowler, T. J., "Acoustic Emission Testing of Vessels and Piping," Chemical Engineering Progress, p. 25, May 1987.
5. Smith, D. T., "Maintenance of Chemical Equipment," from Safety and Accident Prevention in Chemical Operations, 2nd ed., Wiley, New York, 1982.
6. Petersen, H. J. S. and Haastrup, P., "Fault-Tree Evaluation Shows Importance of Testing Instruments and Controls," Chemical Engineering, p. 85, November 26, 1984.
7. Krisher, A. S., "Plant Integrity Programs," Chemical Engineering, p. 20, May 1987.
8. Ostrotsky, B., "Materials Identification in the Plant-I," Chemical Engineering, p. 91, December 15, 1980.
9. Murphy, R., "Computerized Preventive Maintenance," from Maintenance Engineering, 3rd ed., McGraw-Hill, New York, 1977.
10. "What's New in Maintenance Management," Chemical Week, p. 39, August 15, 1984.
11. Conroy, J. A., "A Computer-based Inspection System," Chemical Engineering Progress, p. 39, May 1987.

9

OPERATING PROCEDURES AND TRAINING

Operating procedures describe the intended operation of the facilities in the form of detailed written information. This information is necessary to ensure that the equipment will be operated according to its design in a safe and consistent fashion by all operating personnel, and to minimize failures caused by human error. Proper training is necessary to ensure that operating procedures are understood, and proper supervision is necessary to ensure the procedures are being followed.

9.1 Operating Procedures

Before any equipment is operated, approved operating procedures should be available. Each set of operating procedures should typically contain the following information:

- process description;
- description of equipment;
- description of control philosophy;
- equipment preparation before startup;
- detailed startup, shutdown, and normal operating instructions;
- equipment clean-out and isolation in preparation for maintenance;
- normal operating limits;
- operating data to be recorded on shift logs;
- emergency procedures;
- block diagrams of equipment and simplified diagrams of complex control loops;
- analytical or test procedures to be performed by operators;

- potential hazards of chemicals handled in equipment; and
- personnel safety equipment and safety regulations.

These procedures should be presented in sufficient detail so there is no confusion as to which line, vessel, valve, or instrument is being referenced. They should describe the operation of the equipment according to the design specifications and what action is required when operating parameters deviate from normal limits. All critical operating parameters should be defined clearly; for example, in a table containing the following information:

- process variable (e.g., temperature, pressure, level, flow, composition);
- control instrument identification number;
- location of sensor, readouts, and alarms;
- normal operating limits;
- effect of exceeding limits and required operator action;
- alarm, relief, and shutdown set points; and
- equipment affected by shutdowns.

Procedures are typically written for each major part of the process, whereas instructions are written for each major piece of equipment. The actual instructions, or a condensed version, should be available as a checklist, particularly during startup and shutdown. During normal operations a data log should be maintained to record critical process variables. Even if data are recorded on computer systems or chart recorders, it is useful to keep a manual record so that the operator is kept aware of the status of critical variables and does not rely strictly on his alarms and shutdowns. An important part of avoiding operation outside of normal limits and/or the resulting need to shut down equipment for safety reasons is to be aware of changes occurring in the process early enough to take corrective action to bring the process back into its normal operating range. Logging of process conditions indicated on local instruments also helps ensure that operators periodically go out into the field.

For each piece of equipment, the response to the following emergency conditions should be discussed:

- excursions in critical operating parameters;
- loss of utilities (e.g., electrical power, cooling water, steam, air, inert gas, combustion fuel, scrubbing or neutralizing fluids);
- mechanical failure (e.g., line or vessel rupture, pump failure, seal failure, agitator failure, heat exchanger tube failure);
- instrument failure (e.g., computer, programmable controller, sensors, area monitor, interlock); and
- external events (e.g., fire, explosion, corrosive release, lightning, freezing, tornado, flood, hurricane, earthquake).

Some of these emergency conditions may be addressed in the site emergency response plan, but where these events may have a specific effect on individual equipment they should be addressed in the operating instructions.

Early detection is critical to effective emergency response. All available detection mechanisms should be discussed in the operating procedures, including visual and other sensory observations and area detectors (e.g., toxic gas, combustible gas, smoke, heat, fire, flame). Then all available means to respond to the emergency should be discussed, including the following:

- isolation of release (e.g., automatically activated, remotely activated, and manual isolation valves);
- control of release (e.g., secondary containment, diversion to flare or scrubber, suppression via water, foam, inert gas, Halon, dry chemical, neutralizing media);
- notification of relevant plant functions; and
- personnel protection required.

All operating procedures and instructions should be reviewed and approved by the person(s) responsible for operating safety before equipment is operated. It is important to include operations and maintenance personnel in the writing and/or review process. These people will have to operate and maintain the equipment and can provide valuable comments that will make the procedures easier to understand and to follow correctly. Also, by participating in developing the procedures, they will assume a more active role in promoting their use.

It is important that, after the first startup and initial operation of a unit, the operating procedures and instructions be reviewed and modified where necessary to improve accuracy, clarity, and consistency with the actual performance of equipment. Any changes made at this time, or in subsequent modifications, should have proper review and approval to ensure adherence to original intent, design, and safety issues. Beyond this initial revision, a normal schedule of periodic updating should be developed. Any time that changes are proposed in normal operating limits, alarms, interlocks, reliefs, or equipment and piping configuration, they should be carefully reviewed and approved before implementation. The operating manuals should be promptly revised to incorporate approved changes and operators instructed accordingly. Outdated procedures should be removed from the operating manual, and copies beyond those necessary for process files destroyed.

9.2 Training

It is critical to the success of a training program that the people to be trained have the proper qualifications and aptitudes. Finding such people will require coordination with personnel or human resource groups and, in many instances, with any trade union involved.

Well-written operating instructions can serve as a classroom training guide. In addition to classroom instruction, a good training program should include a substantial amount of time spent in the field tracing lines; identifying all equipment; and learning the location of critical instruments, isolation valves, reliefs, vents, drains, area monitors, and emergency equipment (e.g., fire control equipment, automatic shutdown locations, self-contained breathing apparatus, safety showers).

According to F.O. Kubias[1]: "Studies have shown that when individuals learn a new job they retain the job information best if they follow oral and visual instructions by doing the job themselves and recalling the important points of the job as they proceed." For example, see the statistics below:

Method of Instruction	Three-Day Information Recall
Telling	10%
Showing	20%
Combination of telling and showing	65%

People generally remember:

- 10% of what they read
- 20% of what they hear
- 30% of what they see
- 50% of what they hear and see
- 70% of what they say
- 90% of what they say as they do a task

For new employees or those with inadequate experience in the unit operations required for a new facility, supplemental training is necessary. Some training aids may be obtained from equipment vendors or from commercial sources. These aids provide general information about operation and maintenance of various types of equipment (e.g., pumps, compressors, heat exchangers). For computer-monitored or -controlled systems, vendors now supply training modules that can be run on the process computer. These process simulators are especially useful in emergency training as process upsets can be programmed into the training module and operators can practice emergency response without endangering the facilities.[2] Process simulators can also be developed for processes that are not monitored or controlled by computer. Regardless of the system, some form of emergency response training should be conducted for potential "worst-case" scenarios, including simulated drills.

On-the-job training should also be utilized. In this mode, trainees should work alongside experienced plant personnel and eventually perform all the operating functions under the direct supervision of an experienced plant person before they are left on their own.

As training progresses, written and verbal testing should be conducted to determine the effectiveness of the training program. Certain minimum acceptable standards for each job must be set for operators to achieve before they are allowed to operate the equipment. Testing should include knowledge of procedures as well as field operation of equipment. Upon satisfactory completion of any training, there should be written documentation added to each employee's file. This documentation may also be useful in determining job advancement for individual employees.

In addition to process training, all operating personnel must be given adequate basic safety training. Recently enacted "right-to-know" legislation in the United States mandates training be given to all employees to inform them about the hazards of the chemicals to which they may be exposed. Other training that should be required is listed below:

- plant safety rules;
- personnel protective equipment (e.g., safety glasses, hard hats, respirators, breathing apparatus);
- use of tools;
- emergency procedures (e.g., loss of process or utility flows, instrument failure, mechanical failure);
- emergency response (e.g., fire fighting, spill control, first aid);
- plant safety organization and responsibilities;
- safe work permits (e.g., hot work, vessel entry, line break);
- administrative controls relating to equipment or process changes and the necessity for approvals;
- accident/incident reporting; and
- housekeeping.

A periodic refresher training program should be established to ensure that operating and maintenance personnel maintain the skills necessary to conduct safe and efficient operations. This is particularly true for unusual and rare upset scenarios that could have a significant impact on operation if not responded to in a proper manner. Retraining is always necessary when modifications or changes are made to equipment, procedures, control systems, and the like.

Procedures and training programs should be developed for those contractors who have to work on-site, as well as for visitors. As a minimum, these programs should include plant safety regulations, a description of the potential hazards of chemicals in the areas the non-employees will visit, and emergency evacuation procedures.

References

1. Kubias, F. O., "Tools and Techniques for Chemical Safety Training," Safety and Accident Prevention in Chemical Operations, 2nd., Wiley, New York, 1982.
2. Glaser, D. C., "The PC Simulator," Chemical Engineering Progress, 45, September 1986.

10

EMERGENCY PREPAREDNESS PLANNING

Regardless of how well engineered, constructed, and operated a particular site, there is always a remote chance that something will go wrong. Being prepared for such eventualities can make the difference between a minor incident with limited and controlled consequences and a major disaster with loss of life, widespread media attention, massive court suits, and public outrage. It follows that facilities that handle hazardous materials must undertake a level of emergency planning that is commensurate with both on-site risks to their employees and off-site risks to their surrounding communities.

In the aftermath of several major incidents in recent years, a great deal has been written by both Government and industry organizations on the topic of planning for chemical-related emergencies. This Section will not attempt to reiterate information that is readily available from other sources. Instead, it will merely introduce recent key sources of guidance and point out some of the most common pitfalls and oversights associated with planning for chemical emergencies.

10.1 Government Planning Guides

10.1.1 Publications of the National Response Team

The National Response Team (NRT) is comprised of representatives of 14 federal agencies having major responsibilities in environmental, transportation, and public health matters. It is the national body for planning, preparedness, and response actions related to oil discharges and releases of hazardous substances to the environment. It is also responsible at the federal level for publication of guidance documents for the preparation and implementation of emergency plans for hazardous substances.

Member agencies include the Environmental Protection Agency (Chair), Department of Transportation (U.S. Coast Guard, Vice-Chair), the Federal Emergency Management Agency, and the U.S. Departments of Commerce, Interior, Agriculture, Defense, State, Justice, Transportation (Research and Special Programs Administration), Health and Human Services, Energy, and Labor, together with the Nuclear Regulatory Commission.

The NRT published the "Hazardous Materials Emergency Planning Guide" in March 1987, which provides a detailed overview of efforts required to:

- select and organize a planning team;
- define the tasks of the planning team;
- develop the plan and individual plan elements; and
- appraise the plan and continue planning.

The guide focuses on the needs of public authorities in local communities across the United States. Nevertheless, it contains highly useful information for industrial planning personnel in terms of the basic elements of the planning process. Additionally, it provides insights into those issues of concern to public authorities and the importance of cooperation and coordination of emergency planning activities between the public and private sectors.

10.1.2 Publications of the Federal Emergency Management Agency

FEMA publishes the "Guide for Development of State and Local Emergency Operations Plans," document (CPG 1-8), which provides information to emergency management planners and state and local government officials about FEMA's concept of planning under the Integrated Emergency Management System (IEMS). This system emphasizes integration of planning for all types of hazards that pose a threat to a community, and provides extensive guidance in the coordination, development, review, validation, and revision of emergency operations plans.

The concepts, if not the specific details, of FEMA's guidance are applicable to individual chemical facilities in that many such sites may be subject to a variety of natural and technological hazards. A single emergency plan that provides "umbrella coverage" for a site can often ensure increased efficiency and effectiveness of a planning effort.

10.1.3 Publications of the U.S. Environmental Protection Agency (EPA)

In June 1985, the EPA announced a comprehensive strategy to deal with the problem of airborne toxic agents in the environment; one portion was designed to address the problem of accidental releases of acutely toxic chemicals. This latter initiative resulted in a November 1985 document, "Chemical Emergency Preparedness Program--Interim Guidance," which discusses a host of issues relating to identification and planning for emergencies involving discharges of such substances to the atmosphere. The interim document is being subjected to an intensive review at the time these guidelines are being written.

10.2 Industry and Private Planning Assistance

10.2.1 Publications of the Chemical Manufacturers Association

The Chemical Manufacturers Association (CMA) has established a Community Awareness and Emergency Response Program (CAER) to encourage local plant managers to take the initiative in cooperating with local communities in the development of integrated emergency plans for response to hazardous material incidents. The NRT guidance document cited above notes that knowledgeable chemical industry representatives can be especially helpful during the planning process, and advises community planners to seek out local CMA/CAER program participants. Specifically, the document points out that many chemical plant officials are both willing and able to share equipment and personnel during response operations.

The CMA publishes three documents that could prove considerably useful during the overall planning process:

- "Community Awareness and Emergency Response Program Handbook,"
- "Site Emergency Response Planning," and
- "Community Emergency Response Exercise Program."

These publications are available at nominal cost from the CMA at 2501 M Street, N.W., Washington, D.C. 20037. Further information on publications can be obtained by telephoning (202) 887-1265.

10.2.2 Publications of the Disaster Research Center

A somewhat more formal and in-depth treatment and coverage of emergency planning can be obtained from the publications list of the Disaster Research Center, University of Delaware, Newark, Delaware 19716; telephone no. (302) 451-6618. A small but representative sample of some excellent reports and articles available from the Center includes

- Tierney, K.J., "A Primer for Preparedness for Acute Chemical Emergencies," Book and Monograph #14, 1980.
- Quarantelli, E.L., "Organizational Behavior in Disasters and Implications for Disaster Planning," Report Series #8, 1985.
- Quarantelli, E.L., and Gray, J., "First Responders and Their Initial Behavior in Hazardous Chemical Transportation Accidents," Preliminary Paper #96, 1985.
- Gray, J., "Three Case Studies of Organized Responders to Chemical Disasters," Misc. Report #29, 1981.

- Quarantelli, E.L., "People's Reactions to Emergency Warnings," Article #170, 1983.
- Quarantelli, E.L., et al., "Evacuation Behavior and Problems: Findings and Implications from the Research Literature," Book and Monograph Series #18, 1984.
- Quarantelli, E.L., et al., "Evacuation Behavior: Case Study of the Taft Louisiana Chemical Tank Explosion Incident," Misc. Report #34, 1983.

10.3 Federal Emergency Planning Requirements

Under the Superfund Amendments and Reauthorization Act of 1986, also known as SARA, the Federal Government has set the stage for mandatory emergency response planning requirements for facilities that house more than specified quantities of toxic materials. Of special interest is Title III of the Act, which is known as the "Emergency Planning and Community Right-to-Know Act of 1986."

Title III of SARA requires the establishment of state commissions, planning districts, and local planning committees. The governor of each state is enjoined to appoint a state emergency response commission whose responsibilities are to include designating emergency planning districts; appointing local emergency planning committees for each district; supervising and coordinating the activities of planning committees; reviewing emergency plans; receiving chemical release notifications; and establishing procedures for receiving and processing requests from the public for information about and/or copies of emergency response plans, material safety data sheets, a list of extremely hazardous substances prepared as part of the EPA's original Chemical Emergency Preparedness Program initiative, inventory forms, and toxic chemical release forms.

The formation of emergency planning districts is intended to facilitate the preparation and implementation of emergency plans. Planning districts may be existing political subdivisions or multijurisdictional planning organizations. The local emergency planning committee for each district must include representatives from each of the following groups or organizations:

- elected state and local officials;
- law enforcement, civil defense, firefighting, first aid, health, local environmental, hospital, and transportation personnel;
- broadcast and print media;
- community groups; and
- owners and operators of facilities subject to the requirements of Title III of SARA.

Each emergency planning committee is enjoined to establish procedures for receiving and processing requests from the public for information about and/or copies of emergency response plans, material safety data sheets, and chemical inventory forms. The committee must designate an official to serve as coordinator of information.

Facilities are subject to emergency planning and notification requirements if a substance on the EPA's list of extremely hazardous substances is present at the facility in an amount in excess of the threshold planning quantity (TPQ) for that substance. The owner or operator of each facility subject to these requirements must notify the appropriate state emergency response commission that the facility is subject to the requirements.

Each facility must also notify the appropriate emergency planning committee of a facility representative who will participate in the emergency planning process as a facility emergency coordinator. Upon request, facility owners and operators are to provide the appropriate emergency planning committee with information necessary for developing and implementing the emergency plan for the planning district.

Title III provisions help to ensure that adequate information is available so that the planning committee will know around which facilities to plan. When there is a release of a chemical identified by Title III of SARA, a facility owner or operator, or a transporter of the chemical, must notify the community emergency coordinator for the emergency planning committee for each area likely to be affected by the release, and the state emergency planning commission of any state likely to be affected by the release.

Each emergency planning committee is expected to prepare an emergency plan by October 1988 and then review it annually. The committee also evaluates the need for resources to develop, implement, and exercise the emergency plan, and makes recommendations with respect to additional needed resources and how to provide them. Each emergency plan must include: facilities and transportation routes related to specific chemicals; response procedures of facilities, local emergency and medical personnel; the names of community and facility emergency coordinators; procedures for notifying officials and the public in the event of a release; methods for detecting a release and identifying areas and populations at risk; a description of emergency equipment and facilities in the community and at specified fixed facilities; evacuation plans; training programs; and schedules for response practice drills. The completed plan is to be reviewed by the state emergency response commission and, upon request, may be reviewed by the Federal Regional Response Team, a regional representative of the National Response Team.

To ensure that information is available for developing emergency plans and assisting in response activities, facility owners or operators are required to submit data for specified chemicals to appropriate government agencies.

Part II of the Federal Register of 22 April 1987 presents the Final Rule (40 CFR Parts 300 and 355) containing the Extremely Hazaradous Substances List, Threshold Planning Quantities, and Emergency Planning and Notification Requirements. It is important that any plant subject to the requirements of SARA follow the proposal and promulgation of future federal regulations in this topical area.

10.4 Common Planning Pitfalls and Oversights

All the information and guidance sources cited above provide highly valuable information about emergency preparedness for the chemical industry. Nevertheless, the authors have in several instances, observed that planning personnel, particularly those who have limited experience in the subject area, tend to repeat certain common mistakes. There are advantages, therefore, to listing and discussing some of the more important of these in somewhat of a "top ten" list; however, by no means should this list be considered all-inclusive.

10.4.1 Delegation of Excessive Responsibility to Individuals

Personnel undertaking the emergency planning process will quickly realize that there are numerous action items that require attention in an emergency, and that each individual item may require that several activities be undertaken rapidly. This often results in a tendency for planners to overload certain key individuals with an excessive number of tasks in the initial (and critical) moments of the emergency. These individuals may find it impossible to satisfy all requirements of the emergency plan and, naturally, will set priorities and postpone or ignore some actions. Ultimately, this may leave the company open to accusations of a mismanaged response, or of waiting too long to initiate certain activities, and so forth, which leads to the finding:

- Planning Rule #1: Give thought to what each person with a role in emergency response must do during the course of an emergency and whether he/she will have the time and resources available to carry out assigned duties within the prerequisite amount of time. Remember that the individuals involved may be under considerable stress, particularly in life or death circumstances.

10.4.2 Delayed Notification of Authorities

Public authorities should be given details of an emergency "immediately." At the same time, plant management personnel often do not wish to declare an emergency to the public every time a plant alarm is sounded without first investigating the nature and magnitude of the incident and gaining some knowledge of what happened. Both "sides" have rightful but conflicting goals.

- Planning Rule #2: Ensure that lines of communication and authority within the plant and within the plant's emergency response organization permit the person(s) responsible for notification of outside authorities to alert these authorities about "real" emergencies within moments of their occurrence. Otherwise, consider "alerting" local officials immediately upon sounding of a discharge or spill alarm and maintaining contact with them while the situation is being investigated.

10.4.3 Inadequate Lines of Communication to Local Authorities and the News Media

As soon as a plant experiences an actual emergency situation involving a toxic discharge, and the public is informed of the situation or begins to experience the effects of exposure (even if this means only an unusual odor in the air), there is a good possibility that the plant will be besieged by numerous telephone calls for information by members of the public, news organizations, various Government agencies, and others. Still others may come "knocking" at the front gate, demanding to speak to someone in authority.

- Planning Rule #3: Ensure free and clear lines of communication between the plant's emergency operations center and outside authorities. This may be accomplished by dedicated telephone "hot lines" to public emergency operations centers, installation of compatible radio transceivers, and/or by the prior installation of telephone lines with unlisted numbers distributed only to preselected outside authorities.

- Planning Rule #4: One of the first steps that has to be taken during a plant emergency is controlling access to the plant by "unauthorized personnel." The emergency plan should ensure that plant security personnel are specifically instructed to allow access to public authorities with legitimate reasons for entering the emergency operations center of the plant.

10.4.4 Setting Policy for Intentional Sacrifice of Plant Property

There are times when only "drastic" measures can prevent environmental or public tragedies. For example, there may be instances in which it is less hazardous and inexpensive in the long run to permit some portion of a facility to continue burning rather than to attempt extinguishment by applying large amounts of water (which could become contaminated and run off to local bodies of water). Similarly, where highly toxic gases being released are flammable or combustible, lives may be saved in downwind areas by intentional ignition of the gases via a flare gun or other device. Such decisions are difficult to make under emergency conditions without preconsideration during the planning process.

- Planning Rule #5: Give careful consideration during the planning process to those circumstances under which it may be better to sacrifice plant property to fire than to expose the facility to environmental pollution liabilities or liabilities associated with public deaths or injuries.

10.4.5 Protection of Plant Employees

It is the responsibility of industry under emergency conditions not only to protect the environment and the public, but to protect its own employees as well. The overall planning process must give careful attention to employee alert and notification procedures, counting of "heads" and identification of missing persons, rescue procedures, evacuation of employees from hazardous locations, control of traffic on site during declared emergencies, and care of the injured.

- Planning Rule #6: Industrial emergency preparedness plans should give careful attention to the needs of employees and other on-site personnel both during and in the aftermath of emergency situations.

10.4.6 Protection of Vital Control Rooms

There are plants in the United States in which process control room operators, by virtue of poor control room design procedures, could be among the first to be exposed to high concentrations of highly toxic airborne agents in the event of an emergency. Yet these very same personnel are the first line of defense in regaining control of a "runaway" situation and suitable substitutes may not be immediately available to take their place should they become disabled.

- Planning Rule #7: Reevaluate the protection offered to operators stationed within critical process control rooms. Ensure adequate supplies of protective clothing and breathing air, as well as secure means of voice communications with the plant emergency operations center.

10.4.7 Back-Up and Supplementary Emergency Operations Centers

During an emergency at a chemical facility, personnel must simultaneously act to bring the situation within plant boundaries under control while also providing full support to outside authorities attempting to protect the health and safety of the surrounding community. Normally, a single emergency center is designated as the site of operations, and this site is likely to be a "beehive" of activity.

- Planning Rule #8: Consider the establishment of two separate emergency centers, one from which plant response personnel can work in peace to regain control of the situation or to mitigate consequences, and one from which other personnel can interact as need be with the public, news media, and Government authorities. The two centers should, of course, have secure and open lines of communication. If either center is itself threatened by the emergency situation, the second should be suitably equipped to serve as a back-up location.

10.4.8 Provision of Medical Treatment Information

It can be a mistake to assume that local health care professionals have full knowledge of the best methods available to treat victims of severe chemical exposures given only the name of the substance. It is also not wise to assume that the information available in material safety data sheets (MSDS) suffices for their purposes.

- Planning Rule #9: Ensure that the plant emergency preparedness plan provides for rapid distribution of detailed medical effects and treatment information on potential chemical exposures of interest to surrounding health care facilities. Assign one or more liaison representatives between the plant's medical department staff and these facilities.

10.4.9 Incomplete Planning

Some industrial emergency preparedness plans have had a tendency on occasion to assign long lists of emergency responsibilities to individual departments within the plant organization. This has been problematic in those cases where the departments in question have not undertaken, or been allocated the time and resources needed to undertake, those activities necessary to ensure that they are prepared to meet assigned responsibilities. A situation results in which the plant's emergency preparedness plan "looks" comprehensive, but in fact has serious weaknesses. Such shortcomings can be avoided by conducting periodic drills and/or tests to ensure operational effectiveness of the plan, since these exercises permit identification of plan weaknesses and formulation of necessary improvements.

- Planning Rule #10: Ensure that each plant department with assigned emergency responsibilities submits a written plan of its own showing how it will meet its duties. Append these departmental planning documents to the overall plan for the facility. Conduct regularly scheduled tests and/or drills to ensure that emergency preparedness plans will function as intended.

APPENDIX A

BIBLIOGRAPHY

This bibliography, covering references relevant to the safe handling and storage of HTHMs, lists documents alphabetically within the following categories:

- General
- Mond Index
- Siting and Design
- Storage and Piping
- Transfer Systems and Instrumentation
- Safety and Reliability

General

AIChE--Center for Chemical Process Safety
"Guidelines for Chemical Process Quantitative Risk Assessment Procedures," in press
"Guidelines for Hazard Evaluation Procedures," 1985
"Guidelines for Use of Vapor Cloud Dispersion Models," 1987
"Guidelines for Vapor Release Mitigation - An Overview," 1988

Chemical Manufacturers Association
"Community Awareness and Emergency Response (CAER) Program Handbook"
"Process Safety Management"
"Site Emergency Response Planning"

Kletz, T., "Cheaper, Safer Plants or Wealth and Safety at Work," Institution of Chemical Engineers, Rugby, England, 1984.

Lees, F. P., "Loss Prevention in the Process Industries," Butterworth, Boston and London, 1980.

Mond Index

Lewis, D. J., "The Mond Fire and Explosion Index Applied to Plant Layout and Spacing," 13th Loss Prevention Symposium, American Institute of Chemical Engineers, 1979.

Siting and Design

"A Guide to Safe Warehousing for the European Chemical Industry," European Council of Chemical Manufacturers' Federations (CEFIC), April 1987.

"American National Standard Safety Requirements for the Storage and Handling of Anhydrous Ammonia, ANSI K61.1-1981," American National Standards Institute, Inc., New York.

"Code of Practice for the Large Scale Storage of Fully Refrigerated Anhydrous Ammonia in the United Kingdom," Chemical Industries Association Ltd., London, 1975.

"Code of Practice for the Storage of Anhydrous Ammonia under Pressure in the United Kingdom, Spherical and Cylindrical Vessels," Chemical Industries Association Ltd., London, 1980.

"DOD Ammunition and Explosives Safety Standards," Dept. of Defense, Office of the Assistant Secretary of Defense (Installations & Logistics), Supt. of Documents, Washington, DC, March 1976.

"Flammable and Combustible Liquids Code, NFPA 30," National Fire Protection Association, Quincy, MA, 1984.

"Standard for the Production, Storage and Handling of Liquefied Natural Gas (LNG), NFPA 59A," National Fire Protection Association, Quincy, MA, 1985.

"Storage and Handling of Liquefied Petroleum Gases, NFPA 58," National Fire Protection Association, Quincy, MA, 1986.

"Storage and Handling of Liquefied Petroleum Gases at Utility Gas Plants, NFPA 59," National Fire Protection Association, Quincy. MA, 1984.

Storage and Piping

"ASME Boiler and Pressure Vessel Code, Section VIII, Rules for Construction of Pressure Vessels, Division 1," American Society of Mechanical Engineers, New York, 1983 edition.

"Chemical Plant and Petroleum Refinery Piping, ANSI/ASME B31.3-1984," with Addenda B31.3a-1984 and B31.3b-1986, American Society of Mechanical Engineers, New York, 10017.

"Chlorine, Codes of Practice for Chemicals with Major Hazards," Chemical Industries Association Ltd., London, 1975.

"Code of Practice for the Large Scale Storage of Fully Refrigerated Anhydrous Ammonia in the United Kingdom," Chemical Industries Association Ltd., London, 1975.

"Code of Practice for the Storage of Anhydrous Ammonia Under Pressure in the United States, Spherical and Cylindrical Vessels," Chemical Industries Association Ltd., London, 1980.

Mead, W. J., " The Encyclopedia of Chemical Process Equipment," Reinhold, New York, 1964.

Peng, D. P., "Failure Mode Diagram - A Graphic System of a Fracture Control Program for Pressure Vessels and Tanks," Technology Advances in Engineering and Their Impact on Detection, Diagnosis and Prognosis Methods, Proceedings 36th Meeting,, Mechanical Failures Prevention Group, Cambridge University Press, London and New York, 1982.

"Piping Systems for Dry Chlorine," Chlorine Institute Pamphlet 6, 11th ed., The Chlorine Institute, Inc., New York, July 1985.

Rolfe, S. T., and J. M. Barson, "Fracture and Fatigue Control in Structures: Applications of Fracture Mechanics," Prentice-Hall, New Jersey, 1977.

"Safety Guidelines for the Manufacture of Chlorine," Chlorine Institute Pamphlet 67, 1st ed., The Chlorine Institute, Inc., New York, revision 2, July 1981.

Transfer Systems and Instrumentation

Draft: "Prevention Reference Manual - Chemical Specific, Volume B: Chlorine," Radian Corporation, Austin, Texas, November 1986, DCN 203-024-84-16.

Draft Summary Report: "Technical Support to the South Coast Air Quality Management District - Toxic Chemical Accidental Air Releases," Radian Corp., Austin, Texas, November 1986, DCN-86-203-024-84-13.

"Guide for Pressure Relieving and Depressuring System," API Recommend Practice 521, 2nd ed., p. 44, American Petroleum Institute, September 1982.

Heard, D. B., "Fail-Safe Devices for the Prevention of Hazardous Materials," U.S. Environmental Protection Agency, EPA 600/S2-84-189.

Pamphlet 6: "Piping Systems for Dry Chlorine," The Chlorine Institute, New York, 1986.

Pamphlet 66: "Chlorine Tank Car Loading, Unloading, Air Padding, Hydrostatic Testing," The Chlorine Institute, New York, 1979.

Richmond, D., "Instrumentation for Safe Operation," Safety and Accident Prevention in Chemical Operations, 2nd ed., Wiley, 1982.

"Standards and Practices for Instrumentation," 6th ed., 1980, Edited by L. M. Ferson, Instrument Society of America.

Safety and Reliability

Bush, S. H., "Pressure Vessel Reliability," Journal of Pressure Vessel Technology, February 1975.

"Canvey: an investigation of potential hazards from operations in the Canvey Island/Thurrock area," Health and Safety Executive, HMSO, London, 1978. Contains appendices covering fire statistics for refineries and pressure vessel failures.

"Canvey: a second report," Health and Safety Executive, HMSO, London, 1981. Provides information on various failure rates as well as discussions about ignition probabilities.

"Clean Water Act," 40 CFR 125, NPDES Permits, Storage and Handling of Toxic Materials.

Code of Federal Regulations, Title 29, Parts 1910 and 1926, "Labor and Construction."

Code of Federal Regulations, Title 49, Parts 178 and 179, "Transportation."

Code of Federal Regulations, Title 49, Part 195, "Transportation of Hazardous Liquids by Pipeline."

"Comprehensive Environmental Response Compensation and Liability Act," (CERCLA or Superfund), 40 CFR 280, Underground Hazardous Chemical Tanks.

Engel, J. R., Ed.,"Pressure-Vessel Failure Statistics and Probabilities," Nuclear Safety, Vol. 15, No. 4, July/August 1974.

Green, A. E. and Bourne, A. J., "Reliability Technology," published by Wiley-Interscience, a division of John Wiley & Sons Ltd., London, 1977 (reprinted). Provides data on electronic, pneumatic and hydraulic, and mechanical equipment.

Lancaster, F., "Failures of Boilers and Pressure Vessels: Their Causes and Prevention," International Journal of Pressure Vessels and Piping, Vol. 1, No. 2, 1973.

"Nonelectric Reliability Notebook," (RADC-TR-75-22), AD/A-005 657, January 1975, reproduced by National Technical Information Service. This report contains information on valves, instruments, pumps. hardware, etc.

Petak, W. A. and Atkisson, A. A., "Natural Hazard Risk Assessment and Public Policy," Springer-Verlag, New York, 1982.

"Proceedings of Symposium on Process Industry Hazards - Accidental Release, Assessment, Containment and Control," The Institute of Chemical Engineers, Symposium Series No. 47, September 1976.

"Reactor Safety Study, Appendix III-Failure Data," WASH-1400 (NUREG-75/014), U. S. Nuclear Regulatory Commission, October 1975. Provides data on human errors as well as equipment failures, and is one of the most extensive sources of failure-on-demand estimates.

(Note: Although objections have been raised regarding the WASH-1400 report, the source of disagreement or dissatisfaction is generally the treatment of uncertainty and not the methodology as a whole. The data set collected and compiled for the study is considered to be reliable and accurate, and it is not subject to the same criticisms as the overall study. The failure rates and probabilities given by the report are based on information from operating industrial facilities such as chemical plants, refineries, power plants and manufacturing facilities.)

"Report on Reliability Survey of Industrial Plants, Part I: Reliability of Electrical Equipment and Part III: Causes and Types of Failures of Electrical Equipment, the Methods of Repair, and the Urgency of Repair," IEEE Transactions on Industry Applications, Vol. IA-IO, No. 2, pp. 213-235 and pp. 242-252, March/April 1974. These give information on circuit breakers, transformers, cables, and other electrical equipment.

"Resource Conservation and Recovery Act (RCRA)," 40 CFR 264.191-195. Hazardous Waste Tanks.

"Risk Analysis of Six Potentially Hazardous Industrial Objects in the Rijnmond Area, A Pilot Study," A Report to the Rijnmond Public Authority, report presented by COVO Steering Committee, Reidel, Dordrecht, Holland, 1982. A compilation of data on all types of equipment failure is provided.

Smith, D. J., "Reliability and Maintainability in Perspective," MacMillan, New York, 1985.

Smith, T. A., and Warwick, W. A., "Survey of Defects in Pressure Vessels Built to High Standards of Construction," SRD-R-30, UKAEA Report, 1974.

APPENDIX B

CHEMICAL EXPOSURE INDEX

Robert A. Smith and Gordon R. Miller
The Dow Chemical Company

Introduction

The Chemical Exposure Index (CEI) provides a method of rating the relative potential of acute health hazard to people from possible chemical release incidents. This index is intended to be a simple, empirical rating of the acute toxic hazards from chemicals in a comparable and quantitative manner. It is the basis for establishing both the frequency and depth of review needed to properly evaluate the particular toxic release hazards, prevention, and mitigation in any individual "fixed" or stationary unit.

The CEI formula takes into account five factors that influence the magnitude of exposure from any potential release:

1. Toxicity (Acute Health Hazard Rating)
2. Quantity (Volatilized Portion)
3. Distance (to each area of concern)
4. Molecular Weight (relates to Vapor Density and Vapor Dispersion)
5. Process Variables (Temperature, Pressure, Reactivity)

Each of these factors has been categorized into directly proportional scale numbers to maintain quick and simple calculations by eliminating any non-linear or logarithmic effects. Absolute measures of risk are very difficult to determine, but this system will provide a method of ranking one risk relative to another. It is <u>not</u> intended to define a particular containment as <u>safe</u> or <u>unsafe</u> or <u>acceptable</u> or <u>unacceptable</u>. This procedure will focus attention on the necessary degree of concern and will provide the opportunity for recommendations, improvements, and concurrences from appropriate knowledgeable people. Please note that flammability and explosion hazards are not included in <u>this index</u>.

CHEMICAL EXPOSURE INDEX

PLANT ______________________ LOCATION______________________

CHEMICAL (One per Index) ______________TOTAL QUANTITY IN PLANT ______

LARGEST SINGLE CONTAINMENT ______________ TEMP. OF CONTAINMENT ______

PRESSURE OF CONTAINMENT __________ BOILING POINT ______________

FLASH POINT, °C ______________ MOLECULAR WEIGHT __________

EMERGENCY EXPOSURE PLANNING GUIDELINES (EEPG),
SEE P. B-4 & TABLE B.1, P. B-11: ______________________

I. Complete the Chemical Exposure Index Summary on Page B-4 and show highest CEI to the Public - - - - - - - - - - [ ]

This CEI number determines both the level and frequency of reviews needed. They are as follows:

0 to 100	Internal review - Superintendent and Major Manager review and sign every two years.
100 to 300	Site Review Committee, Superintendent and Major Manager review every two years.
300 to 400	Area Representative and preceding group review annually.
Above 400	Corporate representative and the preceding group review annually.

II. Complete Chemical Exposure Review Checklist on page B-7, prepare Review package, and indicate date ready for further reviews. (Will be scheduled with Loss Prevention Contact) ______

III. List any Sights, Odors, or Sounds which might come from your facility and cause public concern or inquiries (e.g., smoke, large relief valves, odors below hazardous levels such as mercaptans or amines, etc.)

____

____

____

<u>Reviewed by</u>: <u>Date</u>

Plant Supt. or Manager: ______________________ __________

Major or Gen. Manager: ______________________ __________

Site Review Representative:
(If Public CEI is above 100) ______________________ __________

Area Review Representative:
(If Public CEI is above 300) ______________________ __________

Corporate Review Representative:
(If Public CEI is above 400) ______________________ __________

EXPOSURE RISK POTENTIAL SUMMARY

Complete the following Table (See Note 1) by filling in and multiplying the appropriate Risk Factor <u>Scale</u> numbers for each of the major Risk areas as identified in the following formula (see Definition of Scale Numbers on next page):

I.	X	II.	X	III.	X	IV.	X	V.	=	VI.
ACUTE HEALTH HAZARD SCALE NO.		VAPOR QUANTITY SCALE NO.		DISTANCE (RECIPROCAL) SCALE NO.		MOLECULAR WEIGHT SCALE		PROCESS PENALTIES SCALE NO.		CHEMICAL EXPOSURE INDEX

Identification of Containment Vessels being evaluated:
Name or Number (See Notes 2 & 3)

Source #1 ____________________

To (1) Public

(2) Other in-company Facility

(3) Non-company Plant or Business

Source #2 (if needed) ____________

To (1) Public

(2) Other in-company Facility

(3) Non-company Plant or Business

I.	II	III.	IV.	V.	VI. (CEI)

<u>NOTE 1</u> - The distance scale number in column III is the only factor normally expected to change, so column III and the corresponding CEI (column IV) are all that would normally be expected to differ from the first calculation for each source.

<u>NOTE 2</u> - Only different sources of the same chemical should be evaluated on this sheet. Use a separate Chemical Exposure Index for every different chemical.

<u>NOTE 3</u> - Use additional sheets if additional sources with differing potential impact are present.

DEFINITIONS OF SCALE NUMBERS

1. Acute Health Hazard - The acute health hazard is affected by both the concentration at which a material is acutely toxic and the vapor pressure which creates the driving force to volatilize and maintain the material in the atmosphere. For purposes of this scale number, the Emergency Exposure Planning Guide (EEPG) in parts per million is used as the toxicity measurement. Volatility is measured by the vapor pressure in mm Hg @ 25°C up to a maximum of 760. An acute health hazard factor is determined by multiplying the EEPG concentration by 760 and dividing by the vapor pressure. The acute health hazard factors have been assigned scale numbers according to the following ranges:

5 = 0 - 0.99 Health Hazard Factor
4 = 1 - 9.9 Health Hazard Factor
3 = 10 - 99 Health Hazard Factor
2 = 100-999 Health Hazard Factor
1 = 1000-100,000 Health Hazard Factor
0 = >100,000 Health Hazard Factor

The scale number as a function of EEPG and vapor pressure may also be seen graphically on Figure B.1. Table B.1 illustrates a number of materials, properties, and scale numbers.

II. Vapor Quantity - Select the maximum weight of material which would be released as a gas, or would vaporize, during 15 minutes of the "worst case" situation possible. Select the worst case of pressure, temperature, mechanical failure, runaway reactions, contamination, or human failure that is possible, but do not arbitrarily assume 100% of the containment would be released and vaporized in 15 minutes.

4 = >10,000 pounds
3 = 1000 - 10,000 pounds
2 = 100 - 999 pounds
1 = 10 - 99 pounds
0 = <10 pounds

III. Distance - The distance between the point of release and (1) the public, (2) other in-company plants, and (3) other businesses. Public is defined as residences, collecting points of people (schools, churches, shopping centers, hotels), or frequently used public roads.

4 = <1000 feet
3 = 1000 - 5280 feet
2 = 1 mile - 5 miles
1 = 5 miles - 15 miles
0 = >15 miles

IV. Molecular Weight - The density of the vapor is directly related to the molecular weight and inversely affects the rate of dispersion. (Note: in both distance and dispersion ratings, nonlinear or logarithmic functions have been ignored since absolute numbers are not the objective.)

4 = >45 daltons
3 = 34 - 45 daltons
2 = 23 - 33 daltons
1 = 15 - 22 daltons
0 = <15 daltons

V. Process Penalties - The two largest driving forces in generating large amounts of vapor in short periods of time are (1) pressure or over-pressure (e.g., liquids well above their boiling points or high pressure gases) or (2) runaway reactions. Therefore, if either condition exists, a penalty is applied. If the pressure on the system is over 10 psig (1.75 atm. absolute), use 2 for this factor. If the material is, or can become, thermally unstable below 100°C, or is reactive with common materials like air, water, or other possible contaminants, use 2 for this factor. If both factors apply, use 4. If neither apply, use 1.

REVIEW OF PROCEDURE AND CHECKLIST

1. Any plant or facility which requires a further review should notify the site Loss Prevention contact, furnish a copy of this Chemical Exposure Index, and agree to an appropriate review date.

2. At least one week before the review, a Review Package should be sent to each member of the review committee. (These names will be furnished after the review is scheduled.)

3. The Review Package should contain:

   A. A simplified process flowsheet showing containment vessels and quantity.

   B. A chemistry sheet (both desired and undesired reactions).

   C. A plot plan and area map marked with 3 circles, the radii of which equals the distances 1, 10, and 100 times the EEPG concentrations which would result from the quantity that could be released. Contact your Process Engineering group for assistance with the appropriate gas dispersion calculations.

   D. Show all gas monitors, spill detectors, or other devices to detect a release and list test methods and frequency.

   E. List plant responses to "worst case" loss of containment.

   F. List methods of communicating with those in areas which could be affected.

   G. List special things which have been done to minimize the impact, for example, dispersion prediction systems in use, direct line to local law enforcement, community education programs, practice drills, etc.

   H. Submit a completed Containment and Compliance Checklist.

4. At the review, the site representative will collect recommendations and furnish these in a letter.

5. Your response, plans, or actions should be reported to the representative within 30 days and maintained with your Chemical Exposure Index.

CONTAINMENT AND COMPLIANCE CHECKLIST

This checklist will assist the Review Committees in evaluating Credit Factors for the things done to prevent, detect, or control "Loss of Containment." Please "check" those that are complete, operational, or in compliance with known company rules, guidelines or requirements. Please be prepared to cover these and any other specific safeguards designed to reduce the related risk. Any item that cannot be checked should be marked with the percentage of completion. Mark N.A. if not applicable.

Complete		Risk Reducing Credits
________	1.	All pressure vessel and relief device systems properly registered and inspections up to date. (No expansion joints or glass devices.
________	2.	Adequate Emergency Block Valves in place, remotely or automatically operated, and regularly tested.
________	3.	All operational controls and systems designed and routinely tested to "fail-safe."
________	4.	Critical Instrument Program up to date (e.g., redundant high level and temp. alarms, shut-downs, etc.).
________	5.	Operating Discipline complete and up to date.
________	6.	Operating Detectors properly placed and tested regularly.
________	7.	Appropriate engineering specifications properly applied (e.g., lethal service, welded fittings, etc.).
________	8.	Are relief vents on toxic containers designed to minimize atmospheric emissions? How? (circle) Scrubber, Flare, or ____________________.
________	9.	Failure analysis and nondestructive testing carried out where needed (e.g., X-ray, vibration analysis or monitoring, acoustical emission, piping flexibility analysis - hot and cold).
________	10.	Physical barriers in place (for traffic, cranes, etc.).
________	11.	Overdesigned for excess pressure, if needed (e.g., pipelines in certain areas, tank cars, trucks, etc.).
________	12.	All personnel properly trained to understand hazards and emergency responses.
________	13.	Emergency Procedures (relating to this exposure potential) in place and annual drill held.
________	14.	Safety Rules and Safety Standards regularly reviewed and enforced.
________	15.	Loss Prevention Principles and Minimum Requirements appropriately applied.
________	16.	Tech Center Guidelines appropriately incorporated.
________	17.	Reactive Chemical Review within 2 years.

________ 18. Loss Prevention Audit within 2 years.

________ 19. Tech Center Audit within 2 years.

________ 20. All new operations and modifications underwent safety pre-startup audit.

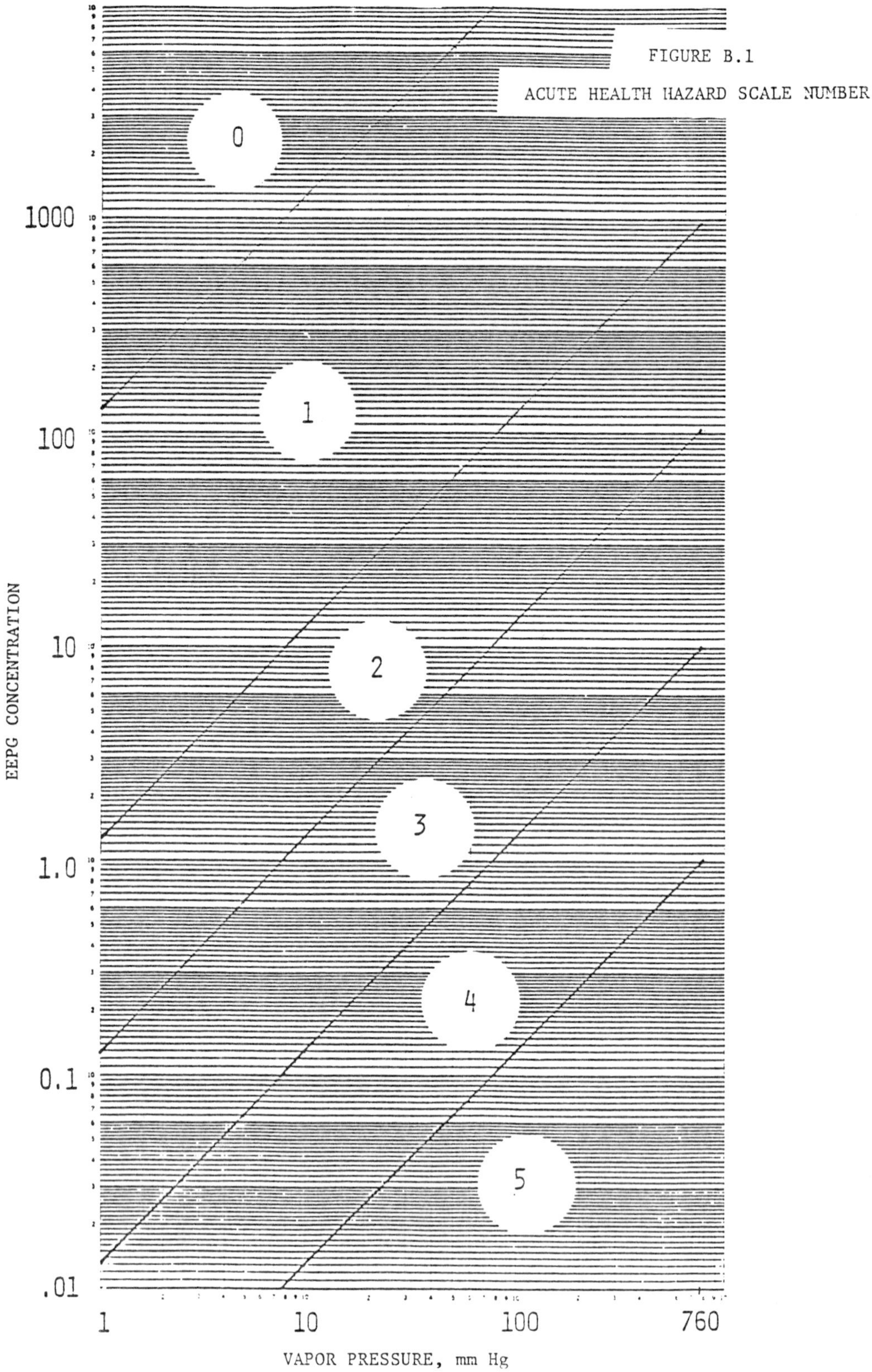
FIGURE B.1
ACUTE HEALTH HAZARD SCALE NUMBER
0
1
2
3
4
5
EEPG CONCENTRATION
1000
100
10
1.0
0.1
.01
1
10
100
760
VAPOR PRESSURE, mm Hg

TABLE B.1

CHEMICAL	EMERGENCY EXPOSURE PLANNING GUIDELINES (EEPG)	MOLECULAR WEIGHT	VAPOR PRES. MM HG @ 25°C	$\frac{760}{\text{VAPOR PRES.}}$	HEALTH* HAZARD FACTOR	ACUTE HEALTH HAZARD SCALE NUMBER (PLACE IN COLUMN I)
Phosgene	0.2	99.	Gas	1	0.2	5
Toluene Diisocyanate	0.2	174.	0.01	76,000	15,200.	1
Chlorine	3.	71.	Gas	1	3.	4
Hydrogen Cyanide	10.	27.	760.	1	10.	3
Sulfur Dioxide	10.	64.	Gas	1	10.	3
Hydrogen Chloride (Anhy.)	25.	36.5	Gas	1	25.	3
Ammonia	100.	17.	Gas	1	100.	2
Acrylonitrile	20.	53.	111.	6.85	137.	2
Ethylene Dichloride	100.	99.	87.	8.74	874.	2
Acetone Cyanohydrin	10.	85.	2.	380	3800.	1**
Bromine	1.	160.	223.	3.40	3.4	4
Hydrogen Bromide	5.	81.	Gas	1	5.	4
Chloroacetyl Chloride	1.	113.	25	30	30.	3

*Health Hazard Factor equals $\frac{\text{EEPG X 760}}{\text{Vapor Pressure}}$

**Dissociates to HCN on heating.

APPENDIX C

TABLE OF CONTENTS

AIChE-CCPS

Guidelines for Vapor Release Mitigation
(in press)

6.3 LIQUID RELEASE COUNTERMEASURES
- 6.3.1 Dilution
- 6.3.2 Neutralization
- 6.3.3 Covers
  - 6.3.3.1 Liquids
  - 6.3.3.2 Foams
  - 6.3.3.3 Solids
  - 6.3.3.4 Application

6.4 AVOIDANCE OF FACTORS THAT AGGRAVATE VAPORIZATION

7. ON-SITE EMERGENCY RESPONSE

7.1 ON-SITE COMMUNICATIONS
7.2 EMERGENCY SHUTDOWN EQUIPMENT AND PROCEDURES
7.3 SITE EVACUATION
7.4 HAVENS
7.5 ESCAPE FROM VAPOR CLOUD
7.6 PERSONAL PROTECTIVE EQUIPMENT
7.7 MEDICAL TREATMENT
7.8 ON-SITE EMERGENCY PLANS, PROCEDURES, TRAINING, AND DRILLS

8. ALERTING LOCAL AUTHORITIES AND THE PUBLIC

8.1 ALERTING SYSTEMS
- 8.1.1 Capabilities
- 8.1.2 Input Requirements

8.2 ROLES AND LINES OF COMMUNICATION
8.3 INFORMATION TO BE COMMUNICATED

9. SELECTION OF MITIGATION MEASURES

9.1 RISK ANALYSIS
9.2 METHODS FOR HAZARD IDENTIFICATION
9.3 METHODS FOR ESTIMATING THE CONSEQUENCES OF ACCIDENTS
9.4 METHODS FOR ESTIMATING THE PROBABILITY OF ACCIDENTS

INDEX